T0325674

Fractional Calculus and Fractional Processes with Applications to Financial Economics

Fractional Calculus and Fractional Processes with Applications to Financial Economics
Theory and Application

Hasan A. Fallahgoul

Sergio M. Focardi

Frank J. Fabozzi

ELSEVIER

AMSTERDAM • BOSTON • HEIDELBERG • LONDON
NEW YORK • OXFORD • PARIS • SAN DIEGO
SAN FRANCISCO • SINGAPORE • SYDNEY • TOKYO
Academic Press is an imprint of Elsevier

Academic Press is an imprint of Elsevier
125 London Wall, London EC2Y 5AS, United Kingdom
525 B Street, Suite 1800, San Diego, CA 92101-4495, United States
50 Hampshire Street, 5th Floor, Cambridge, MA 02139, United States
The Boulevard, Langford Lane, Kidlington, Oxford OX5 1GB, United Kingdom

Notices

Knowledge and best practice in this field are constantly changing. As new research and
experience broaden our understanding, changes in research methods, professional practices, or
medical treatment may become necessary.

Practitioners and researchers must always rely on their own experience and knowledge in
evaluating and using any information, methods, compounds, or experiments described herein.
In using such information or methods they should be mindful of their own safety and the
safety of others, including parties for whom they have a professional responsibility.

To the fullest extent of the law, neither the Publisher nor the authors, contributors, or editors,
assume any liability for any injury and/or damage to persons or property as a matter of
products liability, negligence or otherwise, or from any use or operation of any methods,
products, instructions, or ideas contained in the material herein.

Library of Congress Cataloging-in-Publication Data
A catalog record for this book is available from the Library of Congress

British Library Cataloguing in Publication Data
A catalogue record for this book is available from the British Library

ISBN: 978-0-12-804248-9

For information on all Academic Press publications
visit our website at https://www.elsevier.com/

Working together
to grow libraries in
developing countries

www.elsevier.com • www.bookaid.org

Publisher: Nikki Levy
Acquisition Editor: Glyn Jones
Editorial Project Manager: Anna Valutkevich
Production Project Manager: Poulouse Joseph
Cover Designer: Mark Rogers

Typeset by SPi Global, India

HAF

To my mother Khadijeh and in memory of my father Abobakr Fallahgoul

SMF

To the memory of my parents

FJF

To my family

About the Authors

Hasan A. Fallahgoul is currently a postdoctoral researcher at the Swiss Finance Institute @ EPFL, in the team lead by Professor Loriano Mancini. Prior to this position, he was a postdoctoral researcher at the European Center for Advanced Research in Economics and Statistics (ECARES), Université Libre de Bruxelles, Belgium. His research interests are in financial econometrics, quantitative finance, Lévy processes, and fractional calculus; specializing in heavy-tailed distributions and their applications to finance. Dr. Fallahgoul has published several papers in scientific journals including *Quantitative Finance*, *Applied Mathematics Letters*, and *Journal of Statistical Theory and Practice*. He holds a PhD and MSc in applied mathematics from the K. N. Toosi University of Technology.

Sergio M. Focardi is a professor of finance, director of the Master's course in Investment, Banking and Risk Management, and researcher at the finance group, ESILV EMLV of the Pole Universitaire De Vinci, Paris. He is a founding partner of The Intertek Group, Paris. Professor Focardi holds a degree in electronic engineering from the University of Genoa, Italy, and a PhD in mathematical finance from the University of Karlsruhe, Germany. A member of the Editorial Advisory Board of *The Journal of Portfolio Management*, he has authored numerous articles, monographs, and books.

Frank J. Fabozzi is a professor of finance at EDHEC Business School (Nice, France) and a senior scientific adviser at the EDHEC-Risk Institute. He taught at Yale's School of Management for 17 years and served as a visiting professor at MIT's Sloan School of Management and Princeton University's Department of Operations Research and Financial Engineering. Professor Fabozzi is the editor of *The Journal of Portfolio Management* and an associate editor of several journals, including *Quantitative Finance*. The author of numerous books and articles on quantitative finance, he holds a doctorate in economics from The Graduate Center of the City University of New York.

Contents

Illustrations

Part I

Theory

Fractional calculus and fractional processes: an overview

In this monograph we discuss how fractional calculus and fractional processes are used in financial modeling, finance theory, and economics. We begin by giving an overview of fractional calculus and fractional processes, responding upfront to two important questions:

1. What is the fractional paradigm for both calculus and stochastic processes?
2. Why is the fractional paradigm important in science in general and in finance and economics in particular?

Fractional calculus is a generalization of ordinary calculus. Calculus proved to be a key tool for modern science because it allows the writing of differential equations that link variables and their rates of change. Differential equations ushered in modern theoretical quantitative science. A key reason for the success of differential equations in science is that the description of physical phenomena simplifies *locally*. For example, to understand how temperature propagates in a rod when we apply a source of heat to one of its extremities. While the actual description of the propagation of temperature can be rather complex in function of the source of heat, locally, the heat equation is quite easy: at any point on the rod, the time derivative of the temperature is proportional to the second space derivative: $\frac{\partial T}{\partial t} = k \frac{\partial^2 T}{\partial x^2}$. This is the key for the success of differential equations in science: phenomena simplify locally and one can write a simple, yet extremely powerful, general law.

However, not all problems are local. Many problems in physics, engineering, and economics are of a global nature. Consider, for example, variational problems, where the objective is to optimize a functional, that is, to find an optimum in a space of functions. Though variational problems are not local, it has been a major mathematical success to translate variational problems into integral or differential equations. Still the formalism of fractional calculus is a very useful tool for solving variational problems.

A widely cited example is Abel's integral equation, which solves the problem of the tautochrone. A tautochrone is a curve on a vertical plane such that the time that it takes a point freely sliding on the curve under the sole influence of gravity to reach the lowest point is a constant independent of its starting point. Abel's integral equations can be solved with a fractional derivative.

The problem of the tautochrone is essentially a variational problem. A profound link has been established in several recent papers between the calculus of

http://dx.doi.org/10.1016/B978-0-12-804248-9.50001-2,

variations and fractional calculus, (see, Gorenflo et al. (2001) and Mainardi et al. (2000).). The calculus of variations is one of the main mathematical tools used in science and engineering. Classical dynamics, for example, can be formulated in variational terms as the problem of maximizing Hamiltonian functionals. Control theory is based on variational principles. Fractional calculus is very useful in formulating and solving variational problems. It is the non local characteristic of fractional calculus that makes it interesting (in science and engineering).

Fractional calculus is also useful in computing probability distribution functions with fat tails, widely used in financial risk management. As we will see in Chapters 4 and 5, the solutions of some fractional partial differential equations represent fat-tailed, stable distributions. These distributions cannot be described with closed formulas. Numerical approximations are needed. Fractional calculus provides useful tools for computing numerically fat-tailed stable distributions.

Solving fractional partial differential equations proved to be an important tool in pricing financial contracts such as options. But classical Black-Scholes equations (see Black and Scholes (1973)) fail in many practical applications of option pricing. As we will see in Chapter 6, fractional differential equations provide better approximations to option pricing.

The fractional paradigm applies not only to calculus but also to stochastic processes, used in many applications in finance and economics such as modeling volatility and interest rates and modeling high frequency data. The key features of fractional processes that make them interesting to these disciplines include the following:

1. Long-range memory
2. Path-dependence, non Markovian properties
3. Self-similarity
4. Fractal paths
5. Anomalous diffusion behavior.

The definition and properties of fractional processes look very different from those of fractional calculus; actually, fractional processes such as fractional Brownian motions are defined in terms of long-range dependence, that is, persistent autocorrelation. The fractional nature of these processes appears in some parameters that characterize autocorrelations, namely the so called Hurst-exponent H, which might assume fractional values as opposed to integer values. And another important feature of fractional processes, self-similarity, is also represented through a fractional exponent H in the relationship $\{X_{ct}\} \stackrel{d}{=} \{c^H X_t\}$.

Anomalous diffusions are defined in terms of the growth rate of the diffusion process. The variance of an anomalous diffusion grows as t^α, that is, $var\,(X_t) \propto t^\alpha$, where α is a fractional exponent as opposed to the $\alpha = 1$ integer exponent of normal diffusions. Anomalous diffusions can be represented through fractional Brownian motions or through continuous-time random walks.

What is the relationship between fractional calculus and the properties (persistent correlations, self-similarity, and growth of variance) of fractional processes and anomalous diffusions? There is indeed a deep relationship insofar as the evolution of the probability distribution of fractional processes and anomalous diffusions can be represented as fractional partial differential equations. These fractional partial differential equations are the fractional equivalent of the Fokker-Planck-Kolmogorov equations associated with normal diffusions. However, the non Markovian character of fractional processes and anomalous diffusions does not allow one to write a fractional equivalent of the dual representation of normal diffusions in terms of stochastic differential equations and Fokker-Planck-Kolmogorov partial differential equations.

Let's now briefly consider some areas where fractional processes and anomalous diffusions find applications in finance and economics. Given the state-of-the-art, it now appears clear that fractional processes are applicable to at least *some* financial processes that exhibit long memory. For example, it has been empirically demonstrated that volatility and trading volumes are long-memory processes. Fractional processes have been proposed and used to model these phenomena. Another source of long-memory processes are ultra-high frequency data (tick-by-tick data) and high-frequency trading strategies. Continuous-time random walks and the associated fractional equations can represent the arrival of orders (and prices) and phenomena associated with inter-arrival times.

Fractional processes have also found application in the modeling of interest rates. However, one area that appeared promising – the representation of stock prices and returns as fractional processes with long-range memory – is now being debated: it was discovered that, in continuous time, representing asset prices as fractional Brownian motions leads to arbitrage conditions. This problem has significantly limited the use of fractional processes in finance – at least for the moment.

These and other applications will be described in Chapter 8. Let's now provide an overview of fractional calculus and fractional processes.

1.1 Fractional calculus

Fractional calculus is a generalization and extension of classical calculus. The development of calculus is usually attributed to Gottfried Leibnitz and Isaac Newton, though the two men had a bitter life-long dispute as to which of them actually developed calculus. Leibnitz published his first work on calculus in 1684; Newton made use of calculus in his *Principia* published in 1687. Although some of the key ideas of calculus had been known for centuries to Greek, Arab, and Persian mathematicians, it was Leibnitz and Newton who created a complete framework for calculus and established the fundamental theorem of calculus that links derivatives and integrals.

The derivative of a function, represented as $\frac{df}{dx}$ or f' , is defined as the limit, if it exists, of the difference quotient: $\frac{df}{dx} = \lim_{\Delta x \to 0} \frac{\Delta f}{\Delta x}$, while the (indefinite) integral is defined as the limit of the sum $\int_a^x f(u)\,du = \lim \sum \Delta f \Delta x$, where the sum is extended to all Δx that form a partition of the interval (a, x).

A function $F(x)$ is called an antiderivative (or primitive function) of $f(x)$ if: $\frac{dF(x)}{dx} = f(x)$. The fundamental theorem of calculus states that the indefinite integral $F(x) = \int_a^x f(u) \, du$ is a primitive function of $f(x)$ and that the definite integral can be computed as $\int_a^b f(u) \, du = F(b) - F(a)$, where F is any primitive function of f. Integration is the inverse operation of differentiation.

By recursively applying the operation of taking a derivative (differentiation), we can define derivatives of any order: $D^n(f) = \frac{d}{dx}\left(\frac{d^{n-1}f}{dx^{n-1}}\right)$. We can apply the same iterative process to integration in order to define the integral of n-th order: . $D^{-n}(f) \equiv \int_a^x \int_a^{y_1} \cdots \int_a^{y_{n-1}} f(y_n) \, dy_n dy_{n-1} \cdots dy_1$. Therefore: $\frac{d^n D^m f}{dx^n} = D^{m-n} f, m > n$.

The following integration formula, generally attributed to Cauchy, holds:

$$D^{-n}f \equiv \int_a^x \int_a^{y_1} \cdots \int_a^{y_{n-1}} f(y_n) \, dy_n dy_{n-1} \cdots dy_1 = \frac{1}{(n-1)!} \int_a^x (x-t)^{n-1} f(t) \, dt$$

$$(1.1)$$

Immediately after the introduction of calculus, it became a key research topic for mathematicians. In 1695, a leading mathematician of his time and author of the first French treatise on calculus, le Marquis de l'Hôpital Guillaume François Antoine wrote to Leibnitz asking what would happen if the order of differentiation were a real number instead of an integer. Leibnitz's reply: "It will lead to a paradox, from which one day useful consequences will be drawn." This exchange between Marquis and Leibnitz is generally considered to be the beginning of fractional calculus. However, the actual development of fractional calculus had to wait until 1832. It was then that Joseph Liouville first introduced what is now called the Riemann-Liouville definition of fractional derivative, based on the Riemann-Liouville fractional integral:

$$D^{-\alpha}f \equiv \frac{1}{\Gamma(\alpha)} \int_a^x (x-t)^{\alpha-1} f(t) \, dt.$$

It is interesting to look at the mathematical strategy used in defining fractional derivatives. The starting point is the Cauchy formula

$$D^{-n}f \equiv \int_a^x \int_a^{y_1} \cdots \int_a^{y_{n-1}} f(y_n) \, dy_n dy_{n-1} \cdots dy_1 = \frac{1}{(n-1)!} \int_a^x (x-t)^{n-1} f(t) \, dt$$

that we mentioned above. Using the Gamma function, the Cauchy formula can be generalized to the Riemann-Liouville fractional integral:

$$D^{-\alpha}f \equiv \frac{1}{\Gamma(\alpha)} \int_a^x (x-t)^{\alpha-1} f(t) \, dt. \tag{1.2}$$

This formula is well-defined in ordinary calculus under rather general conditions that will be discussed in the chapters to follow. Therefore, it can be used to generalize the notion of integral of order n, where n is an integer, to that of integral of order $\alpha > 0$ where α is a real number.

The next step is to define the derivative of order $\alpha > 0$. In essence, the process is the following. We know how to define $D^{-\alpha}f$ for any real number and we know how to define $\frac{d^n f}{dx^n}$ for any integer number. To define $\frac{d^\alpha f}{dx^\alpha}$ for any real number $\alpha > 0$, we find the largest integer m such that $m - 1 < \alpha < m$ and we define the (left) fractional derivative of order α in two steps. First we perform a fractional integration of order $m - \alpha$, which is a positive number. Then we take the m-th order derivative. This is called the *left-hand fractional derivative* given by

$$\frac{d^\alpha f}{dx^\alpha} = \frac{d^m}{dx^m}\left(J^\alpha f\right) \tag{1.3}$$

Many other definitions of fractional derivatives have been proposed. Most definitions are based on the Riemann-Liouville integral; other definitions extend the notion of the difference quotient. In this monograph we will discuss several of these definitions. Using fractional derivatives, we can extend the notion of differential equations to fractional differential equations and fractional integral equations.

Let's now ask: What is the geometrical or physical interpretation of fractional calculus? What is the difference between classical calculus and fractional calculus? Why is fractional calculus important? Besides the purely mathematical interest, what is the practical interest of fractional calculus?

The question of the geometrical or physical interpretation of fractional calculus has been discussed for some 300 years without reaching a conclusion. Classical integral and derivatives have well-known and clear geometrical interpretations in terms of tangent lines, tangent planes, and areas. Physical interpretations are also clear in terms of speed and acceleration. However, there is no consensus on the interpretation of fractional integrals and derivatives. The interpretation of fractional derivatives and integrals was recognized as one of the unresolved problems at the first International Conference on Fractional Calculus held in 1974 in New Haven, Connecticut.

Subsequent attempts to provide an interpretation have been made. For example, Podlubny (1998) proposed the "fence interpretation" for stochastic integrals. The possibility of links between fractional calculus and fractals has also been proposed. However, the problem has remained basically unresolved. Despite this, some of the main features and advantages of fractional calculus are well-known, in particular the properties of non locality and the representation of (long) memory processes.

1.2 Fractional processes

After outlining the main concepts of fractional calculus, let's now discuss the fractional paradigm for stochastic processes. As we will see, the fractional paradigm for stochastic processes has two representations:

1. direct description of fractional processes, especially fractional Brownian motion,
2. the fractional equivalent of the Fokker-Planck equation for describing processes such as the continuous-time random walks.

However, we will see that there is no simple duality of representations as in the case of diffusions.

In order to understand the peculiar characteristics of fractional processes and their relationships with fractional calculus, it is useful to outline the classical theory of diffusions. Diffusions have a dual equivalent representation in terms of stochastic differential equations and deterministic partial differential equations. After outlining the classical theory of diffusions, we will see how these representations carry on to fractional processes and abnormal diffusions.

Much neoclassical finance theory and economics are based on Markovian continuous-time stochastic processes. Using a representation which has become standard in finance theory and economics, consider first a probability space $(\Omega, \Im(\Im_t), P)$ formed by a set of possible "states of the world" $\omega \in \Omega$, a sigma-algebra $\Im$ endowed with a filtration $(\Im_t)$, and a probability measure P. A stochastic process is an infinite collection of random variables $X_t(\omega)$ indexed by time. Given t, $X_t(\omega)$ is a random variable; given ω, $X_t(\omega)$ is a path of the stochastic process.

A stochastic process is called *Markovian* if it has the *Markov property* which is expressed as follows: $P(X_s | \Im_t) = P(X_s | X_t), t < s$ or, equivalently, $P(X_s | \Im_t) = P(X_s | \sigma(X_t), t < s)$. That is, the probability distribution of X_s given the information available at time $t<s$, which is embodied in the sigma-algebra $\Im_t$, is equal to the probability distribution of X_s given the value taken by X_t.

Financial and economic processes are often represented as diffusions. A *diffusion* is a Markovian stochastic process with continuous sample paths. Diffusions are characterized by a drift and diffusion coefficients. Two main representations of diffusions are used in finance and economics. The first is based on stochastic differential equations:

$$dX_t = \mu(X_t, t)\, dt + \sigma(X_t, t)\, dB_t \qquad (1.4)$$

where μ, σ are deterministic functions of (X_t, t) called, respectively, the *drift* and the *diffusion coefficients*, and dB_t is the stochastic differential of the standard Brownian motion B_t, also called a *Wiener process*. The *standard Brownian motion* is defined as a process that starts at zero, has continuous paths, and independent Gaussian increments. Paths of Brownian motions have fractal dimension. It can be demonstrated that such a process exists.

The second representation is based on a partial differential equation, the *Forward Kolmogorov equation*, also called the *Fokker-Planck equation*. The Fokker-Planck equation describes the evolution of the conditional probability density function p of X_t. It states that:

$$\frac{\partial p}{\partial t} = -\frac{\partial}{\partial x}\left(\mu(x, t)\, p\right) + \frac{1}{2}\frac{\partial^2}{\partial x^2}\left(\sigma^2(x, t)\, p\right) \qquad (1.5)$$

where p is the conditional density $p \equiv p(x, t | y, s)$ and μ, σ the drift and the diffusion coefficients.

The two representations (1.4) and (1.5) are equivalent under mild mathematical conditions. Historically, the stochastic differential equations approach (1.4) is due to Langevin, who in 1908 established the Langevin equation - a stochastic differential equation; the deterministic partial differential equation approach (1.5) is due to Einstein. The Langevin approach, equation (1.4), was studied mathematically by Itô, who established Itô stochastic calculus. The Einstein approach, equation (1.5), was studied mathematically by Fokker, Planck, and Kolmogorov.

It can be proven that the processes defined by equations (1.4) and (1.5) are diffusion processes. A diffusion process is a Markov process with continuous sample paths, which satisfies the following two conditions:

$$\lim_{t \to s^+} \tfrac{1}{t-s} \int_{|y-x|<\varepsilon} (y-x) \, p\,(x,t\,|y,s)\,dy = \mu(y,s)$$
$$\lim_{t \to s^+} \tfrac{1}{t-s} \int_{|y-x|<\varepsilon} (y-x)^2 \, p\,(x,t\,|y,s)\,dy = \sigma^2(y,s) \qquad (1.6)$$
$$\forall \in> 0, s \geq 0, x \in R$$

The two functions μ, σ^2 in (1.6) are the same as the drift and diffusion coefficients in equations (1.4) and (1.5).

The diffusion equations (1.4) and (1.5) describe many important physical phenomena, including Brownian motion. The Brownian motion is so called in honor of the botanist Robert Brown who, in 1827, described the random motion of small pollen grains suspended in a liquid. Brownian motion is due to random collisions of the pollen grain with the liquid's molecules. In 1905, Einstein introduced the mathematical concepts that led to the Fokker-Planck equation. Einstein used Brownian motion to prove the existence of atoms and molecules - a concept still debated at the beginning of the 20th century. Diffusions are also used to describe, at least approximately, the evolution of stock prices. The French mathematician Louis Bachelier was the first, in 1900, to use diffusions to describe stock prices.

However, many physical processes as well as financial and economic processes cannot be described by the diffusion equations (1.4) and (1.5). In fact, many physical and financial processes are described by what are called *anomalous diffusions*. In the chapters to follow we will discuss many financial processes that can be represented as anomalous diffusions.

Let's now consider the difference between normal and anomalous diffusions, the key one of which is the mean displacement of the particle. If the drift and diffusion coefficients do not depend on time, the variance of a normal diffusion process grows proportionally to time:

$$Var\,[X_t] \propto t \qquad (1.7)$$

An anomalous diffusion is characterized by a non linear growth of the variance which typically follows a power law:

$$Var\,[X_t] \propto t^\alpha. \qquad (1.8)$$

The different behavior of variance in function of time is not a marginal feature. Normal and anomalous diffusions are processes with different characteristics. Normal diffusions are memoryless, Markovian processes; anomalous diffusions are path-dependent, non Markovian processes with long-range dependence. Though all diffusions have continuous, non differentiable sample paths, the fractal dimensions of the paths of normal and anomalous diffusions are different.

Long-range dependence was first studied in hydrology by Hurst (1951). In a monumental work that started in 1906 and lasted 62 years, Hurst documented and described mathematically the long-range dependence properties of the water level of the river Nile. His work was instrumental in the planning of the Aswan High Dam project.

After the invention of fractals, a link was established between long-range memory, fractals, and self-similarity. A process X_t is called *self-similar* if there is a constant H such that, for any scaling factor c, the process X_{ct} and the process $c^H X_t$ have the same distribution. A process is called *H-ss self-similar* if for any constant c there is an H such that $\{X_{ct}\} \overset{d}{=} \{c^H X_t\}$.

Self-similarity has different causes. In 1968, Mandelbrot and Van Ness (1968) described the fractional Brownian motion which is a Gaussian, self-similar process with fractal paths, long-range memory, and stationary increments. However, alpha-stable Lévy processes are self-similar processes with no memory. Therefore, self-similarity does not imply long-range memory. There are also examples, due to Cheridito (2004), which show that long-range memory does not imply self-similarity.

We can now ask: How do we define fractional processes? What is the link between long-range memory processes and fractional calculus? Can we establish a theory of anomalous diffusions similar to that of normal diffusions with dual equivalent descriptions in terms of stochastic differential equations and Fokker-Planck equations?

To discuss these questions, we will look at two processes: fractional Brownian motion and continuous-time random walks. The answer is only partially positive. As we will see we can link continuous-time random walks with fractional partial differential equations. We can also link fractional Brownian motions with fractional partial differential equations. However, due to its non Markovian character, there is no simple way to associate fractional Brownian motions with stochastic differential equations.

A fractional Brownian motion – fBm - B_t^H with Hurst parameter $0<H<1$, is a Gaussian process with continuous paths with covariance:

$$E\left(B_t^H B_s^H\right) = \frac{1}{2}\left(s^{2H} + t^{2H} - |t-s|^{2H}\right)$$

An fBm becomes a standard Bm for $H=0.5$. It is a self-similar process with index H.

A fractional Brownian motion has stationary increments. A process has stationary increments if the distribution of $\{X_{t+h} - X_t\}$ is independent of t for any h. A Brownian motion has independent stationary increments; a fractional Brownian motion has stationary increments which are not independent.

A fractional Brownian motion is an anomalous diffusion with $E[B_t^H]^2 \propto |t|^{2H}$. It is a path-dependent, non Markovian process. A fractional Brownian motion is not a semi-martingale, so the usual Itô stochastic calculus does not apply. That is, it is not possible to define stochastic differential equations that describe a fractional Brownian motion.

It is possible, however, to represent a fractional Brownian motion with a fractional stochastic integral and it is possible to describe the evolution of the probability distribution of a fractional Brownian motion with a fractional equivalent of the Fokker-Planck equation. These constructions will be described in the chapters to follow.

Let's now discuss continuous-time random walks. Following Scalas (2000) a continuous-time random walk is a stochastic process x_t that represents quite naturally the evolution of logprices $x_t = \log(S_t)$. Prices, and therefore logprices, are determined when trades occur. Trades occur at random times t_i. The time series of logprices (x_{t_i}) is characterized by the joint probability $\varphi(\xi, \tau)$ of jumps $\xi_i = x_{t_{i+1}} - x_{t_i}$ and waiting times between jumps $\tau_i = t_{i+1} - t_i$.

Let $p(x, t)$ denote the probability density function of finding a value x_t at time t. Shlesinger and Montroll (1984) demonstrated that the Fourier-Laplace transform $\tilde{p}(k, s)$ of $p(x, t)$ obeys the following equation:

$$\tilde{p}(k, s) = \frac{1 - \tilde{\psi}(s)}{s} \frac{1}{1 - \tilde{\varphi}(k, s)} \tag{1.9}$$

where $\tilde{\psi}(s)$ is the Laplace transform of

$$\psi(\tau) = \int \varphi(\xi, \tau) \, d\xi \text{ and } \tilde{\varphi}(k, s) = \int_0^\infty dt \int_{-\infty}^{+\infty} dx e^{-st+ikx} p(x, t).$$

Suppose now that the distribution of the jump size is independent from the distribution of waiting times: $\varphi(\xi, \tau) = \lambda(\xi) \psi(\tau)$. Under this assumption, equation (1.9) becomes:

$$\tilde{p}(k, s) = \frac{1 - \tilde{\psi}(s)}{s} \frac{1}{1 - \tilde{\lambda}(k) \tilde{\psi}(s)}$$

where $\tilde{\lambda}(k)$ is the Fourier transform of $\lambda(\xi)$.

Let's now make the further assumption that $\tilde{\lambda}(k) = \exp(-|k|^\alpha), 0 < \alpha \leq 2$ and that under these assumptions, $\tilde{\lambda}(k)$ is the characteristic function of an alpha–stable distribution of index alpha. Scalas *et al.* (2000) demonstrate that the following fractional partial differential equation holds:

$$\frac{\partial^\beta p(x, t)}{\partial t^\beta} = \frac{\partial^\alpha p(x, t)}{\partial |x|^\alpha} + \frac{t^{-\beta}}{\Gamma(1 - \beta)} \delta(x), t > 0$$

Continuous-time random walks are non Markovian, long-range memory processes which can be described by the fractional equivalent of the Fokker-Planck equation. Continuous-time random walks do not have continuous paths as paths make jumps. However, the mean squared displacement, that is the variance of the process, does not grow linearly with time. Continuous time random walks have therefore been proposed as a model of anomalous diffusion.

In the following chapters we will discuss in mathematical detail the properties of fractional calculus and fractional processes and their applications in finance and economics.

Fractional Calculus

Since the early 1960s, there have been a good number of papers related to heavy tail distributions. These papers support the view that the heavy tail property is a stylized fact about financial time series. Stable distributions have infinite variance, a property which is not found in empirical samples where empirical variance does not grow with the size of the sample. As an example of the application of fractional calculus, by extending diffusion equations to fractional order, a connection between stable and tempered stable distribution with fractional calculus can be made.

Because there are different methods for extending ordinary calculus to fractional calculus, in this chapter we discuss the concepts and properties of the most well-known definitions of fractional calculus and those that will be used in this monograph.

This chapter is organized as follows. Some important definitions for fractional calculus are discussed in Section 2.1. More specifically, details of the Riemann-Liouville fractional derivative, Caputo fractional derivative, Grünwald-Letnikov fractional derivative, and fractional derivative based on the Fourier transform are discussed. Some examples for implementation of fractional derivatives in Matlab are given in Section 2.2.

2.1 Different definitions for fractional derivatives

The derivative and integral operator are the inverse of each other. More precisely, let D^n be a derivative of order n, then D^{-n} is an integral operator of order n. Figure 2.1 shows the possible choices for the derivative and integral operator in the ordinary calculus.

In ordinary calculus, the possible number for the order of a derivative and integral operator belongs to the integer numbers (such as shown in Figure 2.1), while in fractional calculus the order of a derivative and an integral operator has more flexibility. In fact, the order of a derivative and an integral can be fractional (i.e., the order can be any real number).[1]

There are different approaches for defining a fractional derivative. More specifically, an integral and a derivative operators based on the connection between

---

[1] Even the order of a derivative and an integral operator can be a complex number, see Fallahgoul (2013), Kilbas et al. (2006), Samko et al. (1993) and Podlubny (1998).

http://dx.doi.org/10.1016/B978-0-12-804248-9.50002-4,

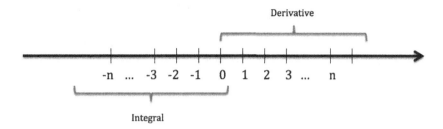

Figure 2.1 Order of derivative or integral operator.

a derivative operator and other operators such as Fourier transform, Laplace transform, n-fold integral operator, finite difference and so on can be extended to upper dimensions. That is, let

$$f_{(n)}(x) = \int_a^x \int_a^{x_1} \cdots \int_a^{x_{n-1}} f(x_{n-1}) dx_{n-1} \cdots dx_1 dx$$

$$= \frac{1}{(n-1)!} \int_a^x (x-t)^{n-1} f(t) dt, \tag{2.1}$$

and

$$f^{(n)}(x) = \frac{d^n}{dx^n} f(x),$$

where $f_{(n)}(x)$ and $f^{(n)}(x)$ are respectively an n-fold integral and a derivative of order n. There are different approaches for presenting the functions $f_{(n)}(x)$ and $f^{(n)}(x)$ in terms of $f(x)$ and n, each approach leading to a definition for the fractional derivative.[2]

Given our objective in this monograph, we limit our discussion to the Riemann-Liouville, Caputo, Grünwald-Letnikov, and fractional integrals and derivatives based on the Fourier transform.[3] Bear in mind that there is a link between these definitions.[4]

2.1.1 Riemann-Liouville Fractional Derivative

In this section, we begin with the definition for the Riemann-Liouville (RL) fractional integrals and derivatives. Using this definition, the RL definition for the fractional derivative is discussed. This definition is more general in terms of containing other definitions. It comes from an n-fold integral operator.

---

[2] See Samko et al. (1993), Kilbas et al. (2006), and Podlubny (1998), among others.
[3] These definitions pass useful properties for practical problem and also other definitions can be obtained from them. See Kilbas et al. (2006) and Podlubny (1998).
[4] More details about these definitions can be found in Fallahgoul (2013), Samko et al. (1993), Kilbas et al. (2006), and Podlubny (1998).

Let $I = [a, b]$ be a finite interval on the real axis $\mathbb{R}$. The left-side and right-side fractional integrals $^{RL}I_{a+}^{\alpha}f$ and $^{RL}I_{b-}^{\alpha}f$ of order $\alpha \in \mathbb{C}$ are defined by

$$(^{RL}I_{a+}^{\alpha}f)(x) = \frac{1}{\Gamma(\alpha)} \int_{a}^{x} \frac{f(t)dt}{(x-t)^{1-\alpha}}, \quad (x > a; Re(\alpha) > 0) \qquad (2.2)$$

and

$$(^{RL}I_{b-}^{\alpha}f)(x) = \frac{1}{\Gamma(\alpha)} \int_{x}^{b} \frac{f(t)dt}{(t-x)^{1-\alpha}}, \quad (x < a; Re(\alpha) > 0) \qquad (2.3)$$

where $\Gamma(x)$ and $Re(\alpha)$ show the gamma function and imaginary part of α, respectively. By setting $n \in \mathbb{N}$ in equations (2.2) and (2.3), we obtain

$$(^{RL}I_{a+}^{\alpha}f)(x) = \frac{1}{(n-1)!} \int_{a}^{x} (x-t)^{n-1}f(t)dt, \quad (n \in \mathbb{N}) \qquad (2.4)$$

and

$$(^{RL}I_{b-}^{\alpha}f)(x) = \frac{1}{(n-1)!} \int_{x}^{b} (t-x)^{n-1}f(t)dt, \quad (n \in \mathbb{N}) \qquad (2.5)$$

It should be stressed that equations (2.4) and (2.5) are equal to equation (2.1).

The left-side and right-side fractional derivatives $^{RL}D_{a+}^{\alpha}f$ and $^{RL}D_{b-}^{\alpha}f$ of order $\alpha \in \mathbb{C}$ are defined by

$$(^{RL}D_{a+}^{\alpha}f)(x) = \left(\frac{d}{dx}\right)^{n} \left(^{RL}I_{a+}^{n-\alpha}f\right)(x)$$

$$= \frac{1}{\Gamma(n-\alpha)} \left(\frac{d}{dx}\right)^{n} \int_{a}^{x} \frac{f(t)dt}{(x-t)^{\alpha-n+1}}, \quad (n = [Re(\alpha)] = 1; x > a)$$

and

$$(^{RL}D_{b-}^{\alpha}f)(x) = \left(-\frac{d}{dx}\right)^{n} \left(^{RL}I_{b-}^{n-\alpha}f\right)(x)$$

$$= \frac{1}{\Gamma(n-\alpha)} \left(-\frac{d}{dx}\right)^{n} \int_{x}^{b} \frac{f(t)dt}{(t-x)^{\alpha-n+1}}, \quad (n = [Re(\alpha)] + 1; x < b)$$

where $[Re(\alpha)]$ shows the integer part of $Re(\alpha)$.

Let $n \in \mathbb{N}$, then $(^{RL}D_{a+}^{0}f)(x) = (^{RL}D_{b-}^{0}f)(x) = f(x)$, $(^{RL}D_{a+}^{n}f)(x) = f^{(n)}(x)$ and $(^{RL}D_{b-}^{n}f)(x) = (-1)^{n}f^{(n)}(x)$.[5]

Remark   Equation $(^{RL}D_{b-}^{0}f)(x) = (-1)^{n}f^{(n)}(x)$ for $n \notin \mathbb{N}$ does not hold.

Example 2.1   Let $f(x) = c$, where c is a constant number, $I = [a, b]$ an arbitrary finite interval on the real axes, and $\alpha = \frac{p}{q} \in (0, 1)$ is a fractional number. The left-side RL fractional derivative of $f(x)$ of order α on interval I is calculated as follows

---

[5] More detailed information about different properties of the RL fractional derivative may be found in Samko et al. (1993) and Kilbas et al. (2006).

$$\left({}^{RL}D_{a+}^{\alpha} c \right)(x) = \left(\frac{d}{dx} \right) \left({}^{RL}I_{a+}^{1-\frac{p}{q}} c \right)(x)$$

$$= \frac{1}{\Gamma(1-\frac{p}{q})} \left(\frac{d}{dx} \right) \int_a^x \frac{cdt}{(x-t)^{\frac{p}{q}}}, \quad (n = [Re(\alpha)] + 1 = 1; x > a)$$

$$= \frac{c}{\Gamma(1-\frac{p}{q})} \left(\frac{d}{dx} \frac{(x-a)^{1-\frac{p}{q}}}{1-\frac{p}{q}} \right),$$

$$= \frac{c(x-a)^{-\frac{p}{q}}}{\Gamma(1-\frac{p}{q})}.$$

Therefore, one important conclusion is that the RL fractional derivative of a constant number is not zero. This property in fractional calculus is in contrast to the property in ordinary calculus where the derivative of a constant function is always zero.

2.1.2 Caputo Fractional Derivative

In this section, a definition for the Caputo (C) fractional derivative is given. The C fractional derivative is defined based on the RL fractional derivative and the Taylor series expansion. If the order of the derivative belongs to $\mathbb{N}$, then the RL and C fractional derivative are equal.

Let $I = [a, b]$ be a finite interval on the real line $\mathbb{R}$, $\alpha \in \mathbb{C}(Re(\alpha) \geq 0)$ and $n = [Re(\alpha)] + 1$. The left-side and right-side C fractional derivatives are defined by

$$\left({}^{C}D_{a+}^{\alpha} f \right)(x) = \frac{1}{\Gamma(n-\alpha)} \int_a^x \frac{f^{(n)}(t)dt}{(x-t)^{\alpha-n+1}} = \left({}^{RL}I_{a+}^{n-\alpha} \left(\frac{d}{dx} \right)^n f \right)(x),$$

and

$$\left({}^{C}D_{b-}^{\alpha} f \right)(x) = \frac{1}{\Gamma(n-\alpha)} \int_x^b \frac{f^{(n)}(t)dt}{(t-x)^{\alpha-n+1}} = (-1)^n \left({}^{RL}I_{b-}^{n-\alpha} \left(\frac{d}{dx} \right)^n f \right)(x),$$

respectively.[6] In particular, when $0 < Re(\alpha) < 1$

$$\left({}^{C}D_{a+}^{\alpha} f \right)(x) = \frac{1}{\Gamma(n-\alpha)} \int_a^x \frac{f'(t)dt}{(x-t)^{\alpha-n+1}} = \left({}^{RL}I_{a+}^{n-\alpha} f' \right)(x),$$

and

$$\left({}^{C}D_{b-}^{\alpha} f \right)(x) = \frac{1}{\Gamma(n-\alpha)} \int_x^b \frac{f'(t)dt}{(t-x)^{\alpha-n+1}} = - \left({}^{RL}I_{b-}^{n-\alpha} f' \right)(x).$$

The left-side and right-side C fractional derivatives can also be defined as

$$\left({}^{C}D_{a+}^{\alpha} f \right)(x) = \left({}^{RL}D_{a+}^{\alpha} \left[f(t) - \sum_{k=0}^{n-1} \frac{f^{(k)}(a)}{k!} (t-a)^k \right] \right)(x),$$

---

[6] In order for this definition to be well-defined, function $f(x)$ must be almost continuous on $[a, b]$. For detailed information about the almost continuous condition, see Samko et al. (1993) and Kilbas et al. (2006), among others.

and

$$\left({}^{C}D_{b-}^{\alpha}f\right)(x) = \left({}^{RL}D_{b-}^{\alpha}\left[f(t) - \sum_{k=0}^{n-1}\frac{f^{(k)}(b)}{k!}(b-t)^{k}\right]\right)(x),$$

respectively.

Both the RL and C fractional derivatives are very common in applied areas. However, in working with differential equations, the C fractional derivative is more flexible. The reason for this attribute for the C financial derivative is related to the initial conditions for solving differential equations: it is not necessary for the initial conditions to be fractional.[7]

Example 2.2   Let $f(x) = x$, $I = [a,b]$ be an arbitrary finite interval on the real axes, and $\alpha = \dfrac{p}{q} \in (0,1)$ is a fractional number. The left-side C fractional derivative of $f(x)$ of order α on interval I is calculated as follows

$$\left({}^{C}D_{a+}^{\alpha}f\right)(x) = \frac{1}{\Gamma(1-\alpha)}\int_{a}^{x}\frac{dt}{(x-t)^{\alpha}}, \quad (n = [Re(\alpha)] = 1; x > a)$$

$$= \frac{1}{\Gamma(1-\frac{p}{q})}\left(\frac{d}{dx}\frac{(x-a)^{1-\frac{p}{q}}}{1-\frac{p}{q}}\right),$$

$$= \frac{(x-a)^{-\frac{p}{q}}}{\Gamma(1-\frac{p}{q})}.$$

2.1.3 Grünwald-Letnikov Fractional Derivative

In this section, we provide another definition for fractional derivative, the Grünwald-Letnikov, named the GL fractional derivative.[8]

The usual definition for differentiation is

$$f^{(n)} = \lim_{h \longrightarrow 0}\frac{(\Delta_{h}^{n}f)(x)}{h^{n}}, \tag{2.6}$$

where $(\Delta_{h}^{n}f)(x)$ is the finite difference of order n for the function $f(x)$ with the step size h, and $n \in \mathbb{N}$.[9] The GL fractional derivative is obtained from equation (2.6) by replacing $\alpha > 0$ instead of n with the finite difference $(\Delta_{h}^{n}f)(x)$ replaced by the fractional order difference $(\Delta_{h}^{\alpha}f)(x)$ which is defined as

$$(\Delta_{h}^{\alpha}f)(x) = \sum_{k=0}^{\infty}(-1)^{k}\binom{\alpha}{k}f(x-kh), \quad (x,h \in \mathbb{R}; \alpha > 0),$$

---

[7] For more detailed information about solving the RL and C fractional differential equation see Hashemiparast and Fallahgoul (2011a) and Hashemiparast and Fallahgoul (2011b), among others.

[8] More detailed information of this definition may be found in Samko et al. (1993) and Kilbas et al. (2006).

[9] For more information about the finite difference, see Samko et al. (1993) and Podlubny (1998).

where

$$\binom{\alpha}{k} = \frac{\alpha(\alpha - 1) \cdots (\alpha - k + 1)}{n!}.$$

The left-side and right-side GL fractional derivatives are defined if $h > 0$ and $h < 0$, respectively. That is, the left-side and right-side GL fractional derivative of order α on finite interval $I = [a, b]$ is defined by

$$\left(^{GL}D^{\alpha}_{a+}f\right)(x) = \lim_{h \longrightarrow 0^+} \frac{\left(\Delta^{\alpha}_{h,a+}f\right)(x)}{h^{\alpha}},$$

and

$$\left(^{GL}D^{\alpha}_{b-}f\right)(x) = \lim_{h \longrightarrow 0^-} \frac{\left(\Delta^{\alpha}_{h,b-}f\right)(x)}{h^{\alpha}},$$

respectively, where

$$\left(\Delta^{\alpha}_{h,a+}f\right)(x) = \sum_{k=0}^{\left[\frac{x-a}{h}\right]} (-1)^k \binom{\alpha}{k} f(x - kh), \quad (x \in \mathbb{R}; h, \alpha > 0),$$

and

$$\left(\Delta^{\alpha}_{h,b-}f\right)(x) = \sum_{k=0}^{\left[\frac{b-x}{h}\right]} (-1)^k \binom{\alpha}{k} f(x + kh), \quad (x \in \mathbb{R}; h, \alpha > 0).$$

2.1.4 Fractional derivative based on the Fourier transform

In this section, a definition for the fractional derivative based on the Fourier transform is provided. This will be extremely useful in solving fractional differential equations, as well as providing a powerful tool for linking the characteristic function of a random variable and a fractional differential equation.[10]

The relation between the Fourier transform and a derivative of order $n (n \in \mathbb{N})$ is as follows

$$\mathbb{F}\left\{\frac{d^n f}{dx^n}\right\} = (i\omega)^n \mathbb{F}\{f\},$$

where $\mathbb{F}$ represents the Fourier transform. By replacing n with a fractional number, i.e., α, a definition for the fractional derivative using the Fourier transform is obtained. In other words, the definition of fractional derivative using the Fourier transform is

$$\left(^F D^{\alpha}_x f\right)(x) = \mathbb{F}^{-1}\left\{(i\omega)^{\alpha} \mathbb{F}\{f\}\right\}, \tag{2.7}$$

and

$$\left(^F D^{\alpha}_{-x} f\right)(x) = \mathbb{F}^{-1}\left\{(-i\omega)^{\alpha} \mathbb{F}\{f\}\right\}, \tag{2.8}$$

---

[10] More detailed information about the connection between fractional differential equations and some distributions is discussed in Fallahgoul et al. (2012b) and Fallahgoul et al. (2013).

where $\mathbb{F}^{-1}$ is the inverse of the Fourier transform. Representing the fractional derivative using Fourier transform by equations (2.7) and (2.8) emphasizes the fact that equation (2.9), in general, for fractional derivatives does not hold; that is,

$$\frac{d^n}{dx^n} = (-1)^n \frac{d^n}{dx^n}. \qquad (2.9)$$

Remark This definition is an extension of the RL fractional derivative for an infinite interval $(-\infty, \infty)$.

2.2 Computation with Matlab

In this section, we discuss implementing the theoretical results for fractional calculus in Matlab.[11]

In order to calculate the RL fractional derivative with Matlab, we will use the mfiles prepared by the Chebfun Team.[12]

We start with ordinary calculus. Command *chebfun(f)* returns the chebfun object for *f*, where *f* can be a string or function handle. Furthermore, command *diff(f)* is used for calculating the first derivative for the function *f*. The output for the following code is presented in Figure 2.2.[13]

```
close all % close all windows;
clear all % reset Workspace history
clc   % clear all command from Command Window
y = chebfun('sin(x)',[-2 2]); % calculate an approximation for sin(x)
y1 = diff(y); % calculate the first derivative
y2 = chebfun('-1*cos(x)',[-2 2]);
LW = 'LineWidth'; lw = 2; % width for the line of the plot
FS = 'FontSize'; fs = 14; % the size of font in title and label for plot
plot(y,'b',y1,'r',y2,'m',LW,lw)    % plot y, y1 and y2 in the same window
legend('sin(x)','cos(x)','-cos(x)'), axis([-2 2 -2 2]), xlabel('x',FS,fs)
title('The function sin(x) with its derivative and antiderivative',FS,fs)
```

In the above derivative, the order of the derivative is 1 which is an integer number. Now, we will obtain the RL fractional derivative for different fractional order. The following codes calculate the RL fractional derivative for different orders $0.1 : 0.01 : 1$.[14] The output of the code is presented in Figure 2.3.

```
y = chebfun('sin(x)',[-8 8]); % calculate an approximation for sin(x)
LW = 'LineWidth'; lw = 2; % width for the line of the plot
FS = 'FontSize'; fs = 14; % the size of font in title and label for plot
```

[11] For more details about using Matlab see, Halpern et al. (2002) and Hunt et al. (2014), among others.
[12] More detailed information about these mfiles may be founded in
   http://www.mathworks.com/matlabcentral/fileexchange/23972-chebfun/content/chebfun/
   examples/integro/html/FracCalc.html
[13] More detailed information regarding the codes are presented in the Matlab code after sign %.
[14] See Matlab help for detailed information about the command $a : d : b$

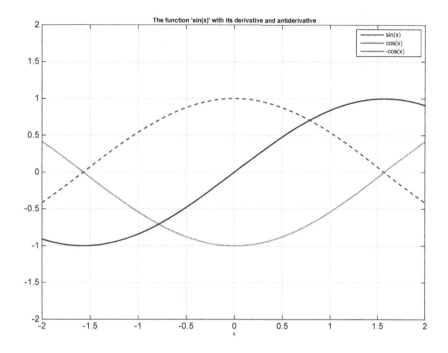

Figure 2.2 The first derivative and integral of sin(x).

```
for alpha = 0.1:.2:1 % the order of RL fractional derivative
    y = [ y diff(y(:,1),alpha) ]; % the RL derivative for order alpha
    plot(y,LW,lw), drawnow
end
xlabel('x',FS,fs); ylabel('d^a sin(x) / d x ^a',FS,fs)
```

By changing the order of the derivative and function, we will get another plot.
Figure 2.4 shows the RL derivatives for $\cos(x) + \sin(x)$.

```
y = chebfun('cos(x)*sin(x)',[-8 8]); % calculate an approximation for sin(x)
LW = 'LineWidth'; lw = 2; % width for the line of the plot
FS = 'FontSize'; fs = 14; % the size of font in title and label for plot
for alpha = 0.1:3:25 % the order of RL fractional derivative
    y = [ y diff(y(:,1),sqrt(2)*alpha/17) ]; % calculating the RL derivative
    plot(y,LW,lw), drawnow
end
xlabel('x',FS,fs); ylabel('d^a cos(x)+sin(x) / d x ^a',FS,fs)
```

By replacing the command *cumsum(f)* rather than *diff(f)* in the above codes,
one can obtain the RL fractional integral.

By comparing Figures 2.2, 2.3, and 2.4, one can see that the difference between a function and its derivative can be substantial while in fractional calculus case this may not be true.

Now, we compute the GL fractional derivative and integral.[15] The output for the following code is presented in Figure 2.5.

```
LW = 'LineWidth'; lw = 2; % width for the line of the plot
FS = 'FontSize'; fs = 14; % the size of font in title and label for plot
h = 0.01; % step size for order of derivative
t = -8:0.1:8;
y = cos(t);
order = 0:0.1:1;
yd     = zeros( length(order), length(t) );
hold on
axis([0,100,-15,15])
for i=1:length(order)
    yd(i,:) = fgl_deriv( order(i), y, h );
    plot(yd(i,:),LW,lw), drawnow
end
```

Key points of the chapter

- There are different ways for extending ordinary calculus to fractional calculus.
- The Riemann-Liouville, Caputo, and Grünwald-Letnikov fractional derivatives are important approaches for extending ordinary calculus to fractional calculus.
- The Riemann-Liouville fractional derivative is the most important extension of ordinary calculus.
- Almost all other definitions for the fractional derivative represent a special case of the Riemann-Liouville fractional derivative.
- In contrast to the ordinary calculus, the fractional derivative of a constant is not zero.
- In working with differential equation, the Caputo fractional derivative is more flexible.
- For constructing a connection between a characteristic function of a random variable and fractional calculus, fractional derivative based on the Fourier transform is more flexible.

---

[15] More detailed information about these mfiles may be found in
http://www.mathworks.com/matlabcentral/fileexchange/45982-fractional-derivative.

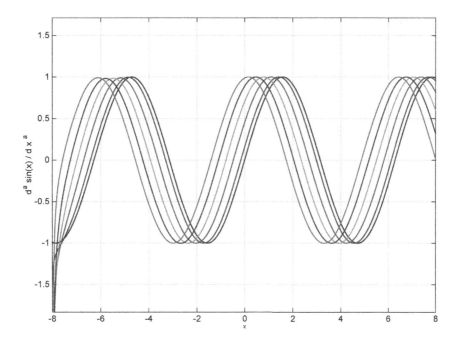

Figure 2.3 The RL fractional derivatives for the $\sin(x)$ of order $a = 0, 0.1, ..., 0.9, 1.0$.

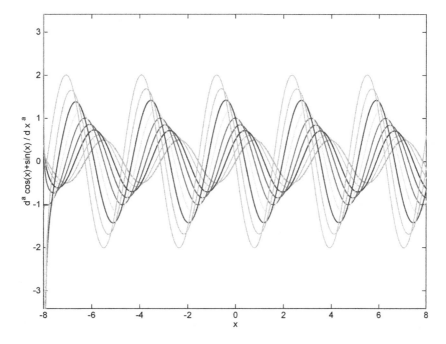

Figure 2.4 The RL fractional derivatives for the $\cos(x) + \sin(x)$ of order $a = \dfrac{\sqrt{2} * (0.1 : 3 : 25)}{17}$.

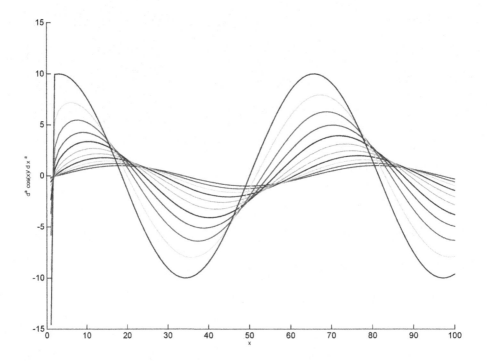

Figure 2.5 The RL fractional derivatives for the $\cos(x)$ of order $a = 0, 0.1, ..., 0.9, 1.0$.

Fractional Brownian Motion

Brownian motion as the source of randomness and uncertainty is used for most applications to real-world problems encountered in financial economics. The Black-Scholes-Merton framework for pricing options is the best known example of the application of Brownian motion.[1] However, there is a preponderance of empirical evidence that fails to support Brownian motion as the source of uncertainty. More specifically, several empirical studies show that the reason for the rejection of the assumption of Brownian motion as the source of randomness is twofold. First, several studies show that asset return distributions observed in financial markets do not follow the Gaussian law because they exhibit excess kurtosis and heavy tails. Second, time series of return distributions exhibit long-range dependency.

In order to deal with the first criticism, several heavy tailed distributions such as the stable, tempered stable, and geo-stable distribution have been proposed to describe return distributions. All of these distributions belong to the class of Lévy processes.[2] To deal with the second criticism, fractional Brownian motion has been used to capture the long-range dependency property of financial data. Fractional Brownian motion, introduced by Mandelbrot and Van Ness (1968), is an extension of Brownian motion just by adding one more parameter which can take on a value that is between zero and one. This parameter is called the Hurst parameter.

While fractional Brownian motion is a useful extension of Brownian motion, there remains one drawback that has been noted in the literature – the possibility of arbitrage. Rogers (1997) proved the possibility of arbitrage, showing that fractional Brownian motion is not a suitable candidate for modeling financial times series of returns. Subsequent to the work by Rogers (1997), there has been considerable debate among researchers about the potential application of fractional Brownian motion in finance. For example, Cheridito (2001b), as well as Rogers (1997), suggested approaches based on changing the weighting kernel of the integral representation for fractional Brownian motion. Alternatively, to eliminate arbitrage, Cheridito (2001a) suggested the following financial asset price dynamic

$$dS_t = \mu S_t dt + \epsilon S_t dB_t + \sigma S_t dB_t^H,$$

where S_t, B_t, and B_t^H are the financial asset price, Brownian motion, and fractional

---

[1] See Black and Scholes (1973) and Merton (1973)

[2] More detailed information about the Lévy process and the distributions discussed above may be found in Cont and Tankov (2003).

http://dx.doi.org/10.1016/B978-0-12-804248-9.50003-6,

Brownian motion, respectively, and μ, σ, and ϵ are the drift, variance, and a small positive number, respectively.

Furthermore, in order to deal with the possibility of arbitrage, some researchers introduced a new stochastic integration such as pathwise and wick-based stochastic integration. Some researchers have criticized these new stochastic integrations, claiming that the financial interpretation for the financial model is changed under these new stochastic integrations.[3]

In this chapter, we discuss the fractional Brownian motion and some of its properties. It should be noted that the fractional concept in this chapter is different from that in the previous two chapters. The structure of the chapter is as follows. The definition of fractional Brownian motion is discussed in Section 3.1. Long-range dependency and self-similarity – two important properties for better understanding fractional Brownian motion – are discussed in Sections 3.2 and 3.3, respectively. In the last section of this chapter, Section 3.4, we show that fractional Brownian motion is not a semimartingale. Indeed, one cannot apply Itô calculus to the fractional Brownian motion.

3.1 Definition

In this section, we discuss fractional Brownian motion (fBm) and its properties. The existence of fBm can be proved through a centered Gaussian process. More precisely, by defining an appropriate covariance function for the Gaussian process one can prove the existence of fBm.

Theorem 3.1 *Let $H > 0$ be a real parameter. Then, there exists a continuous Gaussian process $B^H = (B_t^H)_{t \geq 0}$ with covariance function given by*

$$\Gamma_H(s,t) = \frac{1}{2}\big(s^{2H} + t^{2H} - |t - s|^{2H}\big), \qquad s, t \geq 0$$

if and only if $H \leq 1$.

Proof   See Nourdin (2012). $\square$

The parameter H is called Hurst parameter. It is related to the long-range dependency of the time series. More specifically, by increasing the lag between pairs, the autocorrelation function of the time series decreases.

Definition 3.1 *Let $H \in (0, 1]$. A fBm with Hurst parameter H is a centered continuous Gaussian process $B^H = (B_t^H)_{t \geq 0}$ with covariance function*

$$E\big[B_t^H B_s^H\big] = \frac{1}{2}\big(s^{2H} + t^{2H} - |t - s|^{2H}\big).$$

One can show that when $H = \frac{1}{2}$, then the fBm is just classical Brownian

---

[3] A comprehensive discussion on the use of fractional Brownian motion is provided by Rostek and Schöbel (2013).

motion. Furthermore, when $H \geq \frac{1}{2}$, then the covariance function is given by

$$E\big[B_t^H B_s^H\big] = H(2H - 1) \int_0^t du \int_0^s dv |v - u|^{2H-2}.$$

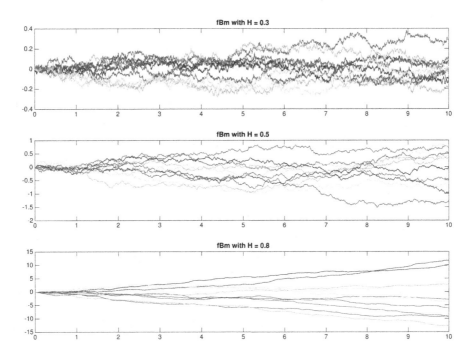

Figure 3.1 10 simulated paths for fBm using different values for the Hurst parameter.

Figure 3.1 shows different simulated paths for fBm. One can see that when the Hurst parameter is near zero, there is less fluctuation. In contrast, for small values for the Hurst parameter there are more fluctuations. When the Hurst parameter is 0.5, the sample paths for fBm are trajectories of Brownian motion.

There are three different integral representations for fBm: time representation, spectral representation, and Volterra process representation.[4]

The time representation of the fBm is given by

$$B_t^H = \frac{1}{C_H} \left(\int_{-\infty}^0 \left((t - u)^{H - \frac{1}{2}} - (-u)^{H - \frac{1}{2}} \right) dB_u + \int_0^t (t - u)^{H - \frac{1}{2}} dB_u \right),$$

---

[4] Detailed information about these representations may be found in Mandelbrot and Van Ness (1968), Decreusefond and Üstüne (1999), Norros et al. (1999), and Nourdin (2012).

where

$$c_H = \sqrt{\frac{1}{2H} + \int_0^\infty \left((1+u)^{H-\frac{1}{2}} - u^{H-\frac{1}{2}}\right)^2 du} < \infty.$$

The spectral representation for the process $B^H = (B_t^H)_{t \geq 0}$ is given by

$$B_t^H = \frac{1}{d_H} \left(\int_{-\infty}^0 \frac{1 - \cos(ut)}{|u|^{H+\frac{1}{2}}} dB_u + \int_0^\infty \frac{\sin(ut)}{|u|^{H+\frac{1}{2}}} dB_u \right),$$

where

$$d_H = \sqrt{2 \int_0^\infty \frac{1 - \cos(ut)}{u^{2H+1}} du} < \infty.$$

The last stochastic integral representation of fBm, the Volterra process, is $B_t^H = \int_0^t K_H(t,s) dB_s$, where B_t is a Brownian motion and K_H is an explicit square integrable kernel.[5]

3.2 Long-Range Dependency

Long-range dependency (LRD) is a measure of decay of statistical dependency. It should be noted that this decay is slower than the decay for an exponential function. From a financial economic prospective, this measure can be an auto-correlation function of lags of a time series. In this section, the definition and properties of the LRD concept are discussed.[6]

Let S_t be a time series for the price of a financial asset and X_t be the log of S_t. Then

$$r_t = X_t - X_{t-1} = \log S_t - \log S_{t-1},$$

is the log-return process of that financial asset.

Figure 3.2 shows the log-price and log-return process for the S&P500 index from March 1, 2000 until December 29, 2010. The log-return process plot reveals that when volatility is high it remains high and when it is low it remains low (volatility clustering property). One of the first approach for modeling this property is using GARCH models. Figure 3.3 shows the sample autocorrelation function of the log-return and absolute log-return process. The sample autocorrelation function for the log-return process reveals that there is not a significant correlation among lags. This is not the case for the sample autocorrelation function for the absolute log-return process where there is positive correlation that decays slowly to zero.

Several researchers show that modeling log-return based on the GARCH approach leads to an exponential decay in autocorrelations of the absolute and

---

[5] Detailed information about this kernel may be found in Nourdin (2012).

[6] There are a considerable number of published papers related to LRD. Bear in mind that LRD is called long memory in the econometrics field. Samorodnitsky (2007) provides a comprehensive discussion about LRD and its properties.

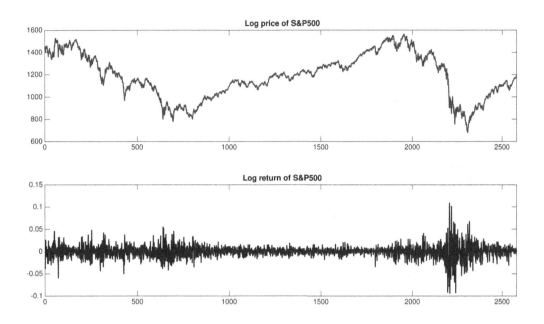

Figure 3.2 Log-price and log-return for the S&P500 index.

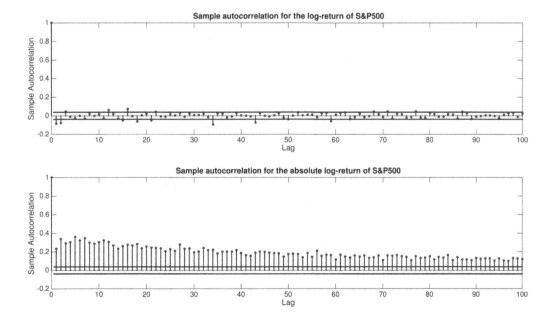

Figure 3.3 Log-price and log-return for the S&P500 index.

squared log-returns.[7] More specifically, the autocorrelation function is similar to the power law and is given by

$$C_{|r|}(\Delta) = corr\left(|r_t|, |r_{t+\Delta}|\right) \simeq \frac{c}{\Delta^\beta},$$

where $\beta \leq 0.5$ and c is a constant.

There are different definitions for the LRD. Guégan (2005), for example, provides 11 definitions of LRD or long memory. The definition for the LRD based on autocorrelation function of a process is related to the slowly varying properties.

Definition 3.2 Let f be a function which for any positive constant a the following equation is satisfied

$$\lim_{x \to \infty} \frac{f(ax)}{f(x)} = g(a) = a^\rho,$$

then f is said to be regularly varying with index ρ. If $\rho = 0$, function f is said to be "slowly varying".

Definition 3.3 A stationary process X_t is said to have LRD if its autocorrelation function decays as a power of the lag. In other words,

$$C(\Delta) = corr\left(X_t, X_{t+\Delta}\right) \simeq \frac{f(\Delta)}{\Delta^{1-2d}}, \qquad 0 < d < \frac{1}{2}$$

where f is slowly varying at infinity.

By using definition 3.3, one can see that the sample autocorrelation function for the absolute or squared log-return process exhibits the LRD property.

One can show that if the Hurst component of fBm exceeds $1/2$, then the covariance of the increments decays very slowly. In other words, the increment of the fBm exhibits the LRD property.

3.3 Self-Similarity

Since the LRD property is based on autocorrelation function for large lags, quantifying this property is not an easy task. Alternatively, self-similar processes can be used for empirical applications instead of quatifying LRD. In this section, a definition for the self-similar process and its relation to LRD are discussed.

Definition 3.4 The stochastic process X_t is said to be a self-similar process if there is a value for H such that for all $c > 0$ one has

$$X_{ct} \stackrel{d}{=} c^H X_t. \tag{3.1}$$

The parameter H is also referred to as the self-similarity exponent, the scaling exponent, and the Hurst exponent.[8]

---

[7] See, among others, Brollerslev et al. (1992), Baillie (1996), and Cont (2005).
[8] Detailed information about self-similar process may be found in Samorodnitsky (2007).

A self-similar process cannot be a stationary process; therefore, it does not process the LRD property. However, if the increment of the process is stationary, then the increment exhibits the LRD property. A fBm is an example of a self-similar process with stationary increments. It should be noted that neither self-similar nor LRD implies the other one. For example, a stable Lévy process is a self-similar process with independent increments.

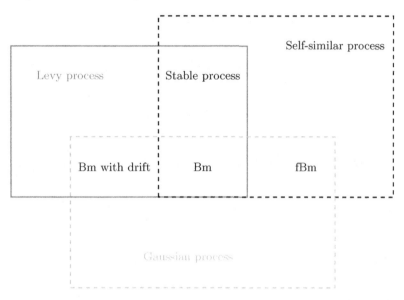

Figure 3.4 The relation of Brownian motion, fBm, Lévy process as well as LRD, and self-similarity property.

Figure 3.4 shows the relation among Brownian motion, fBm, Lévy process, Gaussian process, and the stable process.

In order to better understand the structure of the power law for a self-similar process, one can replace c with $1/t$. More specifically, by setting $c = \frac{1}{t}$ we have

$$X_t \stackrel{d}{=} t^H X_1,$$

therefore

$$F_t(x) = F_1(\frac{x}{t^H}),$$

$$f_t(x) = \frac{1}{t^H} f_1(\frac{x}{t^H}),$$

where F and f are respectively the cumulative distribution function and probability density function for the process X_t. By setting $x = 0$, the probability density function at time t is represented by the scaled probability density function at time 1. That is

$$f_t(0) = \frac{1}{t^H} f_1(\frac{0}{t^H}). \tag{3.2}$$

In general, there are two approaches for checking the self-similar property.

The first is by using equation (3.2) for estimating the Hurst parameter. With this approach, first an estimation for the $f_t(0)$ is obtained using an empirical histogram or kernel estimator and then by using regression to estimate for the Hurst parameter.[9] The second approach is the curve fitting method. This approach is based on comparing the aggregation properties of empirical densities.[10]

3.4 Existence of Arbitrage

The presence of arbitrage is not allowed in modeling asset prices. Therefore, it is important that the underlying asset price process be arbitrage free. On the other hand, it is known that the arbitrage possibilities can be eliminated if and only if the underlying asset price process is a semimartingale. In this section, the arbitrage possibility for fBm is discussed. More specifically, we demonstrate that fBm is not a semimartingale.[11]

In order to show that a fBm is not an arbitrage-free model, it is sufficient to demonstrate that it is not a semimartingale. In fact, the semimartingale property provides the general framework for stochastic integration and the theoretical development in arbitrage pricing. The concept of the semimartingale is based on a cadlag and filtration. We first provide the definitions of a cadlag process and then move on to the concept of filtration.

Definition 3.5 Let $f : [0, T] \longrightarrow \mathbb{R}^n$. f is called a cadlag if it is right-continuous with left limits.

The concept of cadlag is important for studying a jump process. It should be noted that any continuous-time process is a cadlag, but a process with the cadlag property can be a discontinuous process. The jump-diffusion and pure-jump processes are examples of discontinuous cadlag processes. One can easily define a cadlag process. For example, consider a step function having jumps at some point. The value of this step function at each jump point is equal to the value after the jump.

Because a stochastic process is based on time and as time goes on more information is revealed, the status of random variables at each point in time is changing. Indeed, in order to have a precise definition of a stochastic process, we need a definition for revealed information during time. The concept of filtration takes care of the dynamics of information.[12]

Definition 3.6 A filtration or information flow on probability space $(\Omega, \mathcal{F}, P)$ is an increasing family of $\sigma-$algebra. More precisely, let $(\mathcal{F}_t)_{t \in [0,T]}$ be a set of information of a process during the time where $\forall r \leq s, \mathcal{F}_r \subseteq \mathcal{F}_s \subseteq \mathcal{F}$. The family $(\mathcal{F}_t)_{t \in [0,T]}$ is called a filtration.

---

[9] Detailed information about this approach may be found in Mantegna and Stanley (1995).

[10] Cont (2005) and references therein provide detailed information about this approach.

[11] Detailed information about the existence of arbitrage and proofs can be found in Rogers (1997), Cheridito (2001b), Cheridito (2001a), Cheridito (2003), Nourdin (2012), and Rostek and Schöbel (2013).

[12] More information about probability space and $\sigma-$algebras can be found in Cont and Tankov (2003).

Generally speaking, a stochastic process can be predictable or nonpredictable. For understanding the concept of semimartingale it is important to know what is the definition of a nonpredictable stochastic process.

Definition 3.7 A stochastic process $(X_t)_{t \in [0,T]}$ is called a nonpredictable stochastic process with respect to filtration $(\mathcal{F}_t)_{t \in [0,T]}$, if for every $t \in [0,T]$, the value of X_t is revealed at time t.

A nonpredictable stochastic process is called an adapted process on the related filtration.

Definition 3.8 Let $\Pi = \{0 \leq t_1 \leq t_2 \cdots \leq t_n\}$ be a partition of $[0,T]$, and also let $\delta = max(t_i - t_{i-1})_{i=1,2,\cdots,n}$. $p-$variation of the stochastic process $(X_t)_{t \in [0,T]}$ is given by

$$v_p(\Pi) = \lim_{\delta \to 0} \sum_{i=1}^{n} |X_{t_i} - X_{t_{i-1}}|^p.$$

$v_1(\Pi)$ and $v_2(\Pi)$ is called the total variation and realized variance, respectively.[13]

Definition 3.9 A cadlag, adapted stochastic process $(X_t)_{t \in [0,T]}$ is called a semimartingale, for a given filtration $(\mathcal{F}_t)_{t \in [0,T]}$, if it can be decomposed as

$$X_t = X_0 + M_t + A_t,$$

where M_t is local martingale and A_t is an adapted cadlag process with finite-variation.[14]

The concept of a semimartingale provides the largest class of stochastic processes that one can apply Itô calculus.[15] A Brownian motion, a Lévy process, and an adapted continuously differentiable process are examples of the semimartingale. The following properties are for the semimartingale:

- The quadratic variation for every stochastic process with semimartingale property exists (finite).
- Products of semimartingales are semimartingales.
- The linear combination of semimartingales are semimartingales.[16]

The rest of this section shows that fBm is not a semimartingale. More precisely, we show that the total variation and its quadratic variation are not finite except when $H = \frac{1}{2}$. Rogers (1997) shows that fBm is not a semimartingale, therefore one cannot apply Itô calculus to it.

---

[13] It should be noted that the value of $V_p(\Pi)$ is based on its existing limit.

[14] Detailed information about local martingale can be found in Øksendal (2003) and Cont and Tankov (2003).

[15] Itô calculus, is named after Japanese mathematician, Kiyoshi Itô, who suggested the stochastic integration for a stochastic process. He defined an integration by parts formula proving the existence of the stochastic integral. Detailed information about Itô calculus can be found in Cont and Tankov (2003), Shreve (2004), and Wiersema (2008).

[16] There are more properties for the semimartingale. More information about the other properties and the proof of them can be found in Øksendal (2003) and Cont (2005).

Theorem 3.2 *Let $N \sim \mathcal{N}(0,1)$ and let $f : \mathbb{R} \to \mathbb{R}$ be a measurable function such that $E\left[f^2(N)\right] < \infty$. Then*

$$\lim_{n \to \infty} \frac{1}{n} \sum_{k=1}^{n} f\left(B_k^H - B_{k-1}^H\right) \to E[f(N)],$$

where B^H is the fBm with Hurst parameter $H \in (0,1)$.

Proof   See Nourdin (2012).   □

The following corollary is about the p-variation of the fBm.

Corollary 3.1 *Let $p \in [1, \infty)$ and $N \sim \mathcal{N}(0,1)$. Then*

$$\lim_{n \to \infty} \sum_{k=1}^{n} |B_{k/n}^H - B_{(k-1)/n}^H|^p \to 0, \quad if \quad p > \frac{1}{H}$$

$$\lim_{n \to \infty} \sum_{k=1}^{n} |B_{k/n}^H - B_{(k-1)/n}^H|^p \to E[|N|^p], \quad if \quad p = \frac{1}{H}$$

$$\lim_{n \to \infty} \sum_{k=1}^{n} |B_{k/n}^H - B_{(k-1)/n}^H|^p \to \infty, \quad if \quad p < \frac{1}{H}.$$

Proof   See Nourdin (2012).   □

Theorem 3.3 *Let B^H be fBm with Hurst parameter $H \in (0, \frac{1}{2}) \bigcup (\frac{1}{2}, 1)$. Then B^H is not a semimartingale.*

Proof   See Rogers (1997).   □

Key points of the chapter

- There are different representations for fractional Brownian motion.
- Empirical studies show that the time series of financial asset prices exhibit long-range dependency.
- Although the long-range dependency property can be studied based on fractional Brownian motion, quantifying long-range dependency is not an easy task.
- A stochastic process with self-similarity behaves the same when viewed at different scales.
- Self-similar processes are a useful alternative for studying long-range dependency.
- A stochastic process is called a cadlag if it be right-continuous and has left limits.
- A filtration is an indexed set of objects. These objects are information at each index, and by increasing index the size of these objects are increased.
- Fractional Brownian motion is not a semimartingale; more specifically, there is the possibility of the existence of arbitrage for fractional Brownian motion.
- The absence of arbitrage is essential in the pricing of assets and financial derivatives.

Fractional Diffusion and Heavy Tail Distributions: Stable Distribution

In this chapter, our objective is twofold. First, we establish a connection between the stable distributions with fractional calculus. This is accomplished by defining appropriate fractional diffusion equations, the fundamental solution of which provides the PDF for the univariate and multivariate stable distributions.[1] Second, by using some analytic-numerical approaches such as the homotopy perturbation method, the Adomian decomposition method, and the variational iteration method, which are used for solving partial differential equations (PDEs), we obtain some analytic-numerical approximations for the PDF of the univariate and multivariate stable distributions.

The structure of the chapter is as follows. In Section 4.1 the link between the univariate stable distribution using fractional calculus and the analytic-numerical approximation for the PDF of the univariate stable distribution is discussed. The connection between the multivariate stable distribution and fractional PDEs as well as the analytic-numerical approximation for the PDF of the multivariate stable distribution is presented in Section 4.2.

4.1 Univariate Stable Distribution

Applying fractional calculus in this section, we demonstrate the connection between the univariate stable distribution. More specifically, some fractional PDEs are introduced whose fundamental solutions provide the PDF of the univariate stable distributions.[2]

Consider a fractional diffusion PDE given by

$$\frac{\partial u}{\partial t} = -\frac{1+\beta}{2c}\frac{\partial^\alpha}{\partial x^\alpha}u(x,t) - \frac{1-\beta}{2c}\frac{\partial^\alpha}{\partial(-x)^\alpha}u(x,t) + \mu\frac{\partial}{\partial x}u(x,t), \qquad (4.1)$$

where $0 < \alpha \le 2$, $\alpha \ne 1$, $-1 \le \beta \le 1$, and $-\infty < \mu < \infty$, also $c = \cos\frac{\alpha\pi}{2}$ and $s = \sin\frac{\alpha\pi}{2}$. Let $H(\omega, t)$ be the Fourier transform of $u(x,t)$ with respect to t. Then

---

[1] Fractional diffusion equations are an extension of ordinary diffusion equations. This extension can be done by considering fractional derivative either in time or space. In this chapter, we extend ordinary diffusion equations by applying the fractional derivative in space.

[2] Detailed information about the existence of these fractional PDEs may be found in Fallahgoul et al. (2012a) andFallahgoul et al. (2012b).

equation (4.1) converts to the following initial value problem

$$\frac{\partial H}{\partial t} = -\frac{1+\beta}{2c}(i\omega)^\alpha H - \frac{1-\beta}{2c}(-i\omega)^\alpha H + (i\mu\omega)H, \qquad (4.2)$$

where the initial value is $\delta(x)$. If $u(x,0) = \delta(x)$, then $H(\omega,0) = 1$. Therefore the solution of equation (4.2) can be obtained as

$$H(\omega,t) = \exp\{-\frac{1+\beta}{2c}(i\omega)^\alpha t - \frac{1-\beta}{2c}(-i\omega)^\alpha t + (i\mu\omega)\}. \qquad (4.3)$$

One can show that the following fractional PDE is equivalent to equation (4.1)

$$\frac{\partial u}{\partial t} = -\frac{\beta}{c}\frac{\partial^\alpha}{\partial x^\alpha} + (1-\beta)\frac{\partial^\alpha}{\partial |x|^\alpha}u(x,t) + \mu\frac{\partial}{\partial x}u(x,t), \qquad (4.4)$$

where $u(x,0) = u_0(x)$, $-\infty < x < \infty$ and $t > 0$. The Fourier transform for the fundamental solution to (4.4) can be written as

$$H(\omega,t) = \exp\{-|\omega|^\alpha t - i\beta sign(\omega)\tan\frac{\alpha\pi}{2}|\omega|^\alpha t + i\mu\omega t\}. \qquad (4.5)$$

Comparing equation (4.5) to the characteristic function (CF) of the univariate stable distribution, one would find that they are equivalent when $\alpha \neq 1$. Consequently, $u(x,t)$ is the PDF of the univariate stable distribution (i.e., $S_\alpha(t^{\frac{1}{\alpha}}, \beta, \mu t)$).

Now, by using some analytic-numerical approaches namely homotopy perturbation method (HPM), Adomian decomposition method (ADM), and variational iteration method (VIM) some analytic-numerical approximations for the PDF of the univariate stable distribution are obtained.[3]

4.1.1 Homotopy Perturbation Method

Using the HPM, the solution of equation (4.1) is given by

$$v = v_0 + v_1 p^1 + v_2 p^2 + v_3 p^3 + \cdots.$$

where $v_0 = u(x,0) = \delta(x)$. By applying the HPM, one can show that

$$v_1(x,t) = \left(\frac{d_1 + (-1)^\alpha d_2}{2\Gamma(-\alpha)}\right)x^{-\alpha-1} \times t,$$

$$v_2(x,t) = \left(\left(\frac{d_1^2 + (-1)^\alpha d_1 d_2 + d_2^2}{2\Gamma(-2\alpha)}\right)x^{-2\alpha-1} + \left(\frac{d_1 + (-1)^\alpha d_2}{2\Gamma(-\alpha-1)}\right)x^{-\alpha-2}\right)\frac{1}{2}t^2.$$

So we derive the following recurrent relation

$$v_j = \int_0^t \left(d_1\frac{\partial^\alpha v_{n-1}}{\partial x^\alpha} + d_2\frac{\partial^\alpha v_{n-1}}{\partial(-x)^\alpha} + \mu\frac{\partial v_{n-1}}{\partial x}\right)dt,$$

---

[3] It should be noted that these approaches are applied in Fallahgoul et al. (2012a). Detailed information about these methods may be found in He (1999) and the references within Fallahgoul et al. (2012a). One can increase the accuracy of the approximations by applying finite difference method or finite element method. This is an ongoing problem deserving further study.

for $j = 3, 4, 5, \cdots$.

$$u_0(x, t) = v_0(x) = \delta(x),$$

$$u_1(x, t) = v_0 + v_1 = \delta(x) + \left(\frac{d_1 + (-1)^\alpha d_2}{2\Gamma(-\alpha)} \right) x^{-\alpha-1} \times t,$$

$$u_2(x, t) = v_0 + v_1 + v_2 = \delta(x) + \left(\frac{d_1 + (-1)^\alpha d_2}{2\Gamma(-\alpha)} \right) x^{-\alpha-1} \times t$$

$$+ \left(\left(\frac{d_1^2 + (-1)^\alpha d_1 d_2 + d_2^2}{2\Gamma(-2\alpha)} \right) x^{-2\alpha-1} + \left(\frac{d_1 + (-1)^\alpha d_2}{2\Gamma(-\alpha-1)} \right) x^{-\alpha-2} \right) \frac{1}{2} t^2,$$

$$\vdots$$

and so on. In this manner, the rest of the components of the homotopy perturbation solution can be obtained. If $u(x, t) = \lim_{n \to \infty} u_n(x, t)$ and we compute more terms, then we can show that $u(x, t)$ is the PDF for the univariate stable distribution with respect to x; that is, $p(x) = u(x, t) = \lim_{n \to \infty} u_n(x, t) = S_\alpha(t^{\frac{1}{\alpha}}, \beta, \mu t)$ where $p(x)$ is the PDF for the univariate stable distribution.

4.1.2 Adomian Decomposition Method

The recurrence relation for the ADM for equation (4.1) can be constructed as

$$
\begin{aligned}
u_0 &= u(x, 0) = \delta(x), \\
u_{k+1} &= \int_0^t \left(D_1 \frac{\partial^\alpha u_k}{\partial x^\alpha} + D_2 \frac{\partial^\alpha u_k}{\partial(-x)^\alpha} + \mu \frac{\partial u_k}{\partial x} u_k \right) dt, \qquad k = 1, 2, \cdots.
\end{aligned}
$$

where $D_1 = -\dfrac{1 + \beta}{2c}$ and $D_2 = -\dfrac{1 - \beta}{2c}$.
So we derive the following recurrent relation

$$u_j = \int_0^t \left(D_1 \frac{\partial^\alpha u_{j-1}}{\partial x^\alpha} + D_2 \frac{\partial^\alpha u_{j-1}}{\partial(-x)^\alpha} + \mu \frac{\partial u_{j-1}}{\partial x} \right) dt, \qquad (4.6)$$

for $j = 3, 4, 5, \cdots$. Consequently, $p(x) = u(x, t) = \sum_{n=0}^\infty u_n(x, t)$ is an analytic-numerical approximation for the PDF of the univariate stable distribution.

4.1.3 Variational Iteration Method

To solve equation (4.1) by means of the VIM, we set

$$u_{n+1} = u_n + \int_0^t \lambda \left(\frac{\partial u_n}{\partial s} - D_1 \frac{\partial^\alpha v_n}{\partial x^\alpha} - D_2 \frac{\partial^\alpha v_n}{\partial(-x)^\alpha} - \mu \frac{\partial v_n}{\partial x} \right) ds,$$

where $u_0(x,t) = u(x,0) = \delta(x)$. Therefore, the u_1 and u_2 is given by

$$u_1(x,t) = \left(\frac{D_1 + (-1)^\alpha D_2}{2\Gamma(-\alpha)}\right) x^{-\alpha-1} \times t.$$

and

$$u_2(x,t) = \left(\left(\frac{D_1^2 + (-1)^\alpha D_1 D_2 + D_2^2}{2\Gamma(-2\alpha)}\right) x^{-2\alpha-1} + \right.$$
$$\left. + \left(\frac{D_1 + (-1)^\alpha D_2}{2\Gamma(-\alpha-1)}\right) x^{-\alpha-2}\right) \tfrac{1}{2}t^2.$$

respectively.

Consequently, we derive the following recurrent relation

$$u_{n+1}(x,t) = u_n(x,t) - \int_0^t \left(\frac{\partial u_n}{\partial s} - D_1 \frac{\partial^\alpha u_n}{\partial x^\alpha} - D_2 \frac{\partial^\alpha u_n}{\partial(-x)^\alpha} - \mu \frac{\partial u_n}{\partial x}\right) ds.$$

In this manner the rest of the components of the VIM can be obtained. If $u(x,t) = \lim_{n\to\infty} u_n(x,t)$ and we compute more terms, then we can show that $u(x,t)$ is the PDF for the univariate stable distribution with respect to x (i.e., the solution converges to the PDF of the univariate stable distribution).

4.2 Multivariate Stable Distribution

In this section, a multivariate fractional PDE is introduced whose fundamental solution provides the PDF for the multivariate stable distribution.[4]

Let $X(t)$ be the position of a particle at time $t > 0$, and $n-$dimentional Euclidean space $\mathbb{R}^n$. Let $P(\mathbf{x},t)$ denote the density of $X(t)$ where the vector $\mathbf{x} = (x_1, x_2, \cdots, x_n) \in \mathbb{R}^n$. Consider a fractional PDE given by

$$\frac{\partial P(\mathbf{x},t)}{\partial t} = -\langle \mathbf{v}, \nabla P(\mathbf{x},t)\rangle + c\nabla_M^\alpha P(\mathbf{x},t), \qquad (4.7)$$

where $\mathbf{v} \in \mathbb{R}^n$, $c \in \mathbb{R}$, $\nabla = (\frac{\partial}{\partial x_1}, \frac{\partial}{\partial x_2}, \cdots, \frac{\partial}{\partial x_n})$, and ∇_M^α is the general (asymmetric) fractional derivative operator for $0 < \alpha \leq 2$, $\alpha \neq 1$. Also, let the initial condition of equation (4.7) be $P(X(0) = 0) = 1$.

Note that from a physics point of view, $\mathbf{v} \in \mathbb{R}^n$ is the drift coefficient, c descries the spreading rate of the dispersion, and equation (4.7) is the general advection-dispersion equation.

Let $f(x,t): \mathbb{R}^n \times \mathbb{R}^+ \longrightarrow \mathbb{C}$ be a function of $x \in \mathbb{R}6n$ with index $t \in \mathbb{R}^+$ such that $f \in L^1(\mathbb{R}^n)$. The Fourier and inverse Fourier transform is given by

$$\hat{f}(k) = F(f(x)) = \int \exp(-i\langle k,x\rangle) f(x) dx,$$

and

$$f(x) = F^{-1}(\hat{f}(k)) = \frac{1}{(2\pi)^n} \int \exp(i\langle k,x\rangle) \hat{f}(k) dk,$$

---

[4] For more details, see Fallahgoul (2013), Fallahgoul et al. (2014), and Meerschaert et al. (1999).

respectively. Then, we can obtain the following definition for the fractional differential operator according to the Fourier transform

$$\nabla_M^\alpha f(\mathbf{x}) = F^{-1}\left(\left[\int_{\|\theta\|=1} (i\langle \omega, \theta\rangle)^\alpha M(d\theta)\right] \hat{f}(k)\right),$$

where $M(d\theta)$ is an arbitrary probability measure on the unit sphere, and F^{-1} denotes the inverse of the Fourier transform.

Taking the Fourier transform of equation (4.7), we obtain

$$\frac{\partial \hat{P}(k,t)}{\partial t} = -i\langle k, \mathbf{v}\rangle \hat{P}(k,t) + c\left[\int_{\|\theta\|=1} (i\langle \omega, \theta\rangle)^\alpha M(d\theta)\right]\hat{P}(k,t), \qquad (4.8)$$

where $\hat{P}(k,t)$ is the Fourier transform of $P(\mathbf{x},t)$ with respect to $\mathbf{x}$. By applying the initial condition, the solution of equation (4.8) is given by (i.e., initial value problem)

$$\hat{P}(k,t) = \exp\left(-i\langle k, \mathbf{v}t\rangle + ct\int_{\|\theta\|=1} (i\langle \omega, \theta\rangle)^\alpha M(d\theta)\right). \qquad (4.9)$$

Using

$$(i\omega)^\alpha = (\exp(i\frac{\pi}{2})\omega)^\alpha = |\omega|^\alpha \cos(\frac{\alpha\pi}{2})\left(1 + isign(\omega)\tan(\frac{\alpha\pi}{2})\right), \qquad (4.10)$$

the right-hand side of equation (4.9) becomes

$$\exp\left(-i\langle k, \mathbf{v}t\rangle + ct\int_{\|\theta\|=1} |\langle k, \theta\rangle|^\alpha \cos(\frac{\alpha\pi}{2})\left(1 + isign(\langle k, \theta\rangle)\tan(\frac{\alpha\pi}{2})\right) M(d\theta)\right).$$
$$(4.11)$$

If one compares equation (4.11) to the CF of the multivariate stable distribution (see Samorodnitsky and Taqqu (1994)), one would find that they are identical for the case of a stable distribution with $\alpha \neq 1$. Thus the Green's function solution to equation (4.7) yields the entire class of the multivariate stable distribution for $\alpha \neq 1$.

Now, based on the HPM, ADM, and VIM some analytic-numerical approximations for the PDF of the multivariate stable distribution are derived.

4.2.1 Homotopy Perturbation Method

To solve equation (4.7) with initial condition $P(X(0) = 0) = 1$ using the HPM, we construct the following homotopy

$$H(p,V) = (1-p)\left(\frac{\partial V}{\partial t} - \frac{\partial P_0}{\partial t}\right) + p\left(\frac{\partial V}{\partial t} + \langle \mathbf{v}, \nabla V\rangle - c\nabla_M^\alpha\right) = 0. \quad (4.12)$$

Suppose the solution of equation (4.7) has the form

$$V(\mathbf{x},t) = p^0 V_0 + p^1 V_1 + p^2 V_2 + \cdots + p^{n-1} V_{n-1} + \cdots. \qquad (4.13)$$

Substituting ((4.13) into equation (4.12), and comparing coefficients of terms with identical powers of p, leads to

$$p^0 : \quad \frac{\partial V_0}{\partial t} - \frac{\partial P_0}{\partial t} = 0,$$

$$p^1 : \quad \frac{\partial V_1}{\partial t} = -\langle \mathbf{v}, \nabla V_0 \rangle + c\nabla_M^\alpha V_0,$$

$$\vdots$$

$$p^{n+1} : \quad \frac{\partial V_{n+1}}{\partial t} = -\langle \mathbf{v}, \nabla V_n \rangle + c\nabla_M^\alpha V_n.$$

For simplicity, we take $V_0(\mathbf{x}, t) = P_0(\mathbf{x}, t)$. By compared coefficients, the following recurrent relation is derived

$$V_0(\mathbf{x}, t) = P_0(\mathbf{x}, t),$$

$$V_1(\mathbf{x}, t) = \int_0^t (-\langle \mathbf{v}, \nabla V_0 \rangle + c\nabla_M^\alpha V_0) \, dt, \qquad V_1(\mathbf{x}, 0) = 0,$$

$$\vdots$$

$$V_{n+1}(\mathbf{x}, t) = \int_0^t (-\langle \mathbf{v}, \nabla V_n \rangle + c\nabla_M^\alpha V_n) \, dt, \qquad V_{n+1}(\mathbf{x}, 0) = 0.$$

The solution is

$$\begin{aligned} P(\mathbf{x}, t) &= \lim_{p \to 1} V(\mathbf{x}, t) \\ &= V_0(\mathbf{x}, t) + V_1(\mathbf{x}, t) + V_2(\mathbf{x}, t) + \cdots + V_{n+1}(\mathbf{x}, t) + \cdots, \\ &= \delta(\mathbf{x}) + \int_0^t (-\langle \mathbf{v}, \nabla V_0 \rangle + c\nabla_M^\alpha V_0) \, dt \\ &\quad + \int_0^t (-\langle \mathbf{v}, \nabla V_1 \rangle + c\nabla_M^\alpha V_1) \, dt + \cdots, \\ &\quad + \int_0^t (-\langle \mathbf{v}, \nabla V_n \rangle + c\nabla_M^\alpha V_n) \, dt + \cdots. \end{aligned} \tag{4.14}$$

Therefore,

$$P(\mathbf{x}, t) = \delta(\mathbf{x}) + \sum_{k=0}^\infty \left[\int_0^t (-\langle \mathbf{v}, \nabla V_k \rangle + c\nabla_M^\alpha V_k) \, dt \right]. \tag{4.15}$$

If $n = 1$ (one-dimensional case) in equation (4.15), we get

$$u(x, t) = \lim_{n \to \infty} u_n(x, t) = \delta(x) + \sum_{k=0}^\infty \int_0^t \left(d_1 \frac{\partial^\alpha v_k}{\partial x^\alpha} + d_2 \frac{\partial^\alpha v_k}{\partial (-x)^\alpha} + \mu \frac{\partial v_k}{\partial x} \right) dt, \tag{4.16}$$

equation (4.16) appears quite similar to the series representations for the stable density.[5]

---

[5] See Feller (2008).

4.2.2 Adomian Decomposition Method

We will solve equation (4.7) with initial condition $P(X(0) = 0) = 1$, using the ADM. To do so, we construct the following recurrence relation

$$V_0(\mathbf{x}, t) = P_0(\mathbf{x}, t) = \delta(\mathbf{x}),$$

$$V_{k+1}(\mathbf{x}, t) = \int_0^t \left(-\langle \mathbf{v}, \nabla V_k \rangle + c \nabla_M^\alpha V_k \right) dt, \qquad k \geq 0.$$

The solution is then obtained as:

$$V_1(\mathbf{x}, t) = \int_0^t \left(-\langle \mathbf{v}, \nabla V_0 \rangle + c \nabla_M^\alpha V_0 \right) dt,$$

$$V_2(\mathbf{x}, t) = \int_0^t \left(-\langle \mathbf{v}, \nabla V_1 \rangle + c \nabla_M^\alpha V_1 \right) dt,$$

$$\vdots$$

$$V_{n+1}(\mathbf{x}, t) = \int_0^t \left(-\langle \mathbf{v}, \nabla V_n \rangle + c \nabla_M^\alpha V_n \right) dt,$$

$$\vdots$$

Therefore,

$$P(\mathbf{x}, t) = \sum_{k=0}^\infty V_k(\mathbf{x}, t) = \delta(\mathbf{x}) + \sum_{k=0}^\infty \left[\int_0^t \left(-\langle \mathbf{v}, \nabla V_k \rangle + c \nabla_M^\alpha V_k \right) dt \right].$$

Therefore, $P(\mathbf{x}, t)$ is derived via an infinite series. Bear in mind that $P(\mathbf{x}, t)$ is the PDF for the multivariate stable distribution.

4.2.3 Variational Iteration Method

To solve equation (4.7)) using VIM, we set

$$V_{n+1}(\mathbf{x}, t) = V_n(\mathbf{x}, t) + \lambda \int_0^t \left(\frac{\partial V_n(\mathbf{x}, s)}{\partial s} + \langle \mathbf{v}, \nabla V_n(\mathbf{x}, s) \rangle - c \nabla_M^\alpha V_n(\mathbf{x}, s) \right) ds, \tag{4.17}$$

so

$$\delta V_{n+1}(\mathbf{x}, t) = \delta V_n(\mathbf{x}, t) + \delta \int_0^t \lambda \left(\frac{\partial P_n(\mathbf{x}, s)}{\partial s} + \langle \mathbf{v}, \nabla P_n(\mathbf{x}, s) \rangle - c \nabla_M^\alpha P_n(\mathbf{x}, s) \right) ds = 0,$$

the stationary conditions of equation (4.17) are:

$$1 + \lambda = 0, \qquad \lambda' = 0. \tag{4.18}$$

The Lagrange multiplier turns out to be $\lambda = -1$.

Substituting $\lambda = -1$ into equation (4.17), we get the following variational iteration formula

$$V_{n+1}(\mathbf{x}, t) = V_n(\mathbf{x}, t) - \int_0^t \left(\frac{\partial V_n(\mathbf{x}, s)}{\partial s} + \langle \mathbf{v}, \nabla V_n(\mathbf{x}, s) \rangle - c\nabla_M^\alpha V_n(\mathbf{x}, s) \right) ds,$$

where $V_0(\mathbf{x}, t) = P_0(\mathbf{x}, t) = \delta(\mathbf{x})$. Therefore,

$$
\begin{aligned}
P(\mathbf{x}, t) &= \lim_{n\to\infty} V_n(\mathbf{x}, t), \\
&= V_0(\mathbf{x}, t) + \sum_{k=0}^{\infty} \left[\int_0^t \left(-\langle \mathbf{v}, \nabla V_k \rangle + c\nabla_M^\alpha V_k \right) dt \right], \\
&= \delta(\mathbf{x}) + \sum_{k=0}^{\infty} \left[\int_0^t \left(-\langle \mathbf{v}, \nabla V_k \rangle + c\nabla_M^\alpha V_k \right) dt \right].
\end{aligned}
\tag{4.19}
$$

Consequently, the rest of the components of the VIM can be obtained. If $\lim_{n\to\infty} V_n(\mathbf{x}, t)$ and we compute more terms, then we can show that $P_1(\mathbf{x}, t)$ is the multivariate stable distribution's PDF with respect to $\mathbf{x}$.

Key points of the chapter

- Stable and tempered stable distributions are useful for modeling heavy tail models.
- Stable distributions do not have a closed-form expression for either their probability density function or cumulative distribution function.
- The probability density function for the stable distributions is the fundamental solution of some fractional diffusion equation for the univariate and multivariate cases.
- The Homotopy perturbation method, Adomian decomposition method, and variational iteration method are useful methods for obtaining the analytic-numerical approximation for some nonlinear problems.
- By applying the Homotopy perturbation method, Adomian decomposition method, and variational iteration method one can obtain a good analytic-numerical approximations for the probability density function of the univariate and multivariate stable distributions.
- The analytic-numerical approximations for the probability density function of the univariate and multivariate stable distribution have singularity in a small neighborhood of zero.

Fractional Diffusion and Heavy Tail Distributions: Geo-Stable Distribution

Geometric stable (geo-stable) distributions are suitable alternatives for the normal distribution, which suffers from lack of heavy-tail and asymmetric property. These distributions share the same problem with stable distributions, which do not have a closed-form formula for either their probability density function (PDF) and cumulative distribution function (CDF). Therefore, the application of these distributions has some difficulties. Fractional calculus provides suitable tools for obtaining the analytic-numerical approximation for the PDF of these distributions allowing for their application.

Let G be a geometrics random variable with mean $\frac{1}{p}$, and $X_1, X_2, \cdots$ be a sequence of independent and identically distributed random variables. $X_1, X_2, \cdots$ are independent of G. If there exist deterministic $a > 0$ and $b \in \mathbb{R}$ such that

$$a \sum_{i=0}^{G} (X_i + b) \longrightarrow Y, \quad \text{as} \ \ p \longrightarrow 0,$$

then Y has a geo-stable distribution in limit.

The structure of the chapter is as follows. In Section 5.1 the link between the univariate geo-stable distribution using fractional calculus as well as their analytic-numerical approximation for the PDF is discussed. The connection between the multivariate geo-stable distribution and fractional PDEs is presented in Section 5.2.

5.1 Univariate Geo-stable Distribution

The geo-stable distributions, introduced by Klebanov et al. (1984), is particularly appropriate in modeling heavy-tail distributions when the variable of interest may be thought of as a result of a random number of independent innovations.[1]

In this section, we show the connection between the univariate and multivariate geo-stable distributions by once again applying fractional calculus as well as some analytic-numerical approximations for their PDF. For $0 < \alpha < 3, C \neq 0$, consider a fractional partial differential equation

$$\frac{\partial u}{\partial t} = C \frac{\partial^\alpha}{\partial x^\alpha} u(x, t), \qquad x \in \mathbb{R}, t > 0. \tag{5.1}$$

---

[1] Detailed information about the geo-stable distribution my be found in Kozubowski and Rachev (1994).

http://dx.doi.org/10.1016/B978-0-12-804248-9.50005-X,

where $u(x, 0) = u_0(x)$.[2]

Let $\hat{u}(\omega, t)$ be the Fourier transform of $u(x, t)$ with respect to x, then we have

$$\frac{\partial \hat{u}}{\partial t} = C(i\omega)^\alpha \hat{u}. \tag{5.2}$$

This can be viewed as an ordinary differential equation with independent variable t. It can be solved as

$$\hat{u}(\omega, t) = \exp(C(i\omega)^\alpha t)\hat{u}(\omega, 0), \tag{5.3}$$

where $\hat{u}(\omega, 0)$ is the Fourier transform of the initial value $u_0(x) = u(x, 0)$.

The inverse Fourier transform of (5.3), called the fundamental solution for equation (5.1), is

$$
\begin{aligned}
u(x, t) &= F^{-1}\Big\{ \exp(C(i\omega)^\alpha t)\hat{u}(\omega, 0) \Big\} \\
&= F^{-1}\Big\{ \exp(C(i\omega)^\alpha t) \Big\} * F^{-1}\Big\{ \hat{u}(\omega, 0) \Big\} \\
&= K(x, t) * u_0(x),
\end{aligned} \tag{5.4}
$$

where $K(x, t)$ is the solution of (5.1) if $u_0(x) = \delta(x)$.

Theorem 5.1   *The fundamental solution $K(x, t)$ of equation (5.1) is the density of the stable distribution $S_\alpha((-Ct\cos(\frac{\alpha\pi}{2}))^{\frac{1}{\alpha}}, 1, 0)$. Note that we assume $1 < \alpha \leq 2$.*

Proof   See Fallahgoul et al. (2012a).   □

Theorem 5.2   *The fundamental solution $K_1(x, t)$ of the equation*

$$\frac{\partial u_1}{\partial t} = u_1(x, t) * C\frac{\partial^\alpha u_1}{\partial x^\alpha}, \qquad x \in \mathbb{R}, \quad t > 0, \tag{5.5}$$

with the initial condition $u_1(x, 0) = \delta(x)$, is the density of the geo-stable distribution, i.e., $GS_\alpha((-Ct\cos(\frac{\alpha\pi}{2}))^{\frac{1}{\alpha}}, 1, 0)$. Note that we assume $1 < \alpha \leq 2$ and the notation "$*$" denotes the convolution operator.

Proof   See Fallahgoul et al. (2012a).   □

Now we define a fractional PDE by

$$\frac{\partial u_1}{\partial t} = u_1(x, t) * \left(-\frac{1+\beta}{2c}\frac{\partial^\alpha}{\partial x^\alpha}u_1(x, t) - \frac{1-\beta}{2c}\frac{\partial^\alpha}{\partial(-x)^\alpha}u_1(x, t) + \mu\frac{\partial}{\partial x}u_1(x, t) \right), \tag{5.6}$$

where $0 < \alpha \leq 2$, $\alpha \neq 1$, $-1 \leq \beta \leq 1$, and $-\infty < \mu < \infty$, also $c = \cos\frac{\alpha\pi}{2}$ and $s = \sin\frac{\alpha\pi}{2}$.

---

[2] Bear in mind that this PDE is the fractional diffusion equation.

If $H_1(\omega, t)$ be the Fourier transform of $u(x, t)$ with respect to t, then equation (5.6) converts to the following IVP

$$\frac{\partial H_1}{\partial t} = H_1(\omega, t) \times \left(-\frac{1+\beta}{2c}(i\omega)^\alpha H_1(\omega, t) - \frac{1-\beta}{2c}(-i\omega)^\alpha H_1(\omega, t) + (i\mu\omega)H_1(\omega, t) \right),$$
(5.7)

where the initial value is $\delta(x)$. If $u(x, 0) = \delta(x)$, then $H_1(\omega, 0) = 1$.

Theorem 5.3 *The fundamental solution $K_1(x, t)$ of equation (5.6) is the density of the geo-stable distribution, i.e., $GS_\alpha(t^{\frac{1}{\alpha}}, \beta, \mu t)$, for $\alpha \neq 1$.*

Proof See Fallahgoul et al. (2012a). □

By using HPM, ADM, and VIM some analytic-numerical approximations for the univariate geo-stable distribution's PDF are obtained.

5.1.1 Homotopy Perturbation Method

Consider the fractional partial differential equation

$$\frac{\partial u}{\partial t} = u(x, t) * C\frac{\partial^\alpha u}{\partial x^\alpha}, \qquad x \in \mathbb{R}, \quad t > 0,$$
(5.8)

subject to initial condition $u(x, 0) = \delta(x)$, and C is a positive coefficient. To solve equation(5.8) with initial condition $u(x, 0) = \delta(x)$, using HPM, the following homotopy is given

$$H(v, p) = (1 - p)\left(\frac{\partial v}{\partial t} - \frac{\partial u_0}{\partial t} \right) + p\left(\frac{\partial v}{\partial t} - v(x, t) * C\frac{\partial^\alpha v}{\partial x^\alpha} \right) = 0,$$
(5.9)

Suppose the solution of equation (5.8) has the form

$$v = v_0 + v_1 p^1 + v_2 p^2 + v_3 p^3 + \cdots,$$
(5.10)

Substituting (5.10) into (5.8), and comparing coefficients of terms with identical powers of p, leads to

$$p^0 : \frac{\partial v_0}{\partial t} - \frac{\partial u_0}{\partial t} = 0$$

$$p^1 : \frac{\partial v_1}{\partial t} = v_0(x, t) * C\frac{\partial^\alpha v_0}{\partial x^\alpha} - \frac{\partial u_0}{\partial t}, \qquad v_1(x, 0) = 0$$

$$p^2 : \frac{\partial v_2}{\partial t} = v_0(x, t) * C\frac{\partial^\alpha v_1}{\partial x^\alpha} + v_1(x, t) * C\frac{\partial^\alpha v_0}{\partial x^\alpha}, \qquad v_2(x, 0) = 0,$$

$$p^3 : \frac{\partial v_3}{\partial t} = \sum_{i=0}^{2}\sum_{j=0}^{2} p^{i+j} v_i(x, t) * C\frac{\partial^\alpha v_j}{\partial x^\alpha}, \qquad v_3(x, 0) = 0, \quad i + j = 2$$

$$\vdots$$

$$p^{n+1} : \frac{\partial v_{n+1}}{\partial t} = \sum_{i=0}^{n}\sum_{j=0}^{n} p^{i+j} v_i(x, t) * C\frac{\partial^\alpha v_j}{\partial x^\alpha}, \qquad v_n(x, 0) = 0, \quad i + j = n.$$

For simplicity we take $v_0(x, t) = u_0(x, t) = \delta(x)$. So we derive

$$
\begin{aligned}
v_1 &= \int_0^t \left(v_0(x, t) * C\frac{\partial^\alpha v_0}{\partial x^\alpha} - \frac{\partial u_0}{\partial t} \right) dt \\
&= \int_0^t \left(v_0(x, t) * C\frac{\partial^\alpha v_0}{\partial x^\alpha} \right) dt, \\
&= \int_0^t \left(\int_0^s \left(v_0(s - x, t) \times C\frac{\partial^\alpha}{\partial x^\alpha} v_0(x, t) \right) dx \right) dt, \\
&= \int_0^t \left(\int_0^s \left(\delta(s - x) \times C\frac{\partial^\alpha}{\partial x^\alpha} \delta(x) \right) dx \right) dt, \\
&= \int_0^t \left(\int_0^s \left(\delta(x - s) \times C\frac{x^{-\alpha-1}}{2\Gamma(-\alpha)} \right) dx \right) dt, \\
&= \int_0^t \left(C\frac{s^{-\alpha-1}}{2^2\Gamma(-\alpha)} \right) dt, \\
&= C\frac{s^{-\alpha-1}}{2^2\Gamma(-\alpha)} \times t,
\end{aligned}
$$

by using change of variable, we have

$$
v_1(x, t) = C\frac{x^{-\alpha-1}}{2^2\Gamma(-\alpha)} \times t,
$$

$$
\begin{aligned}
v_2 &= \int_0^t \left(v_0(x, t) * C\frac{\partial^\alpha v_1}{\partial x^\alpha} + v_1(x, t) * C\frac{\partial^\alpha v_0}{\partial x^\alpha} \right) dt \\
&= \int_0^t \left(\int_0^s \left(v_0(s - x, t) \times C\frac{\partial^\alpha}{\partial x^\alpha} v_1(x, t) \right) dx \right) dt, \\
&+ \int_0^t \left(\int_0^s \left(v_1(s - x, t) \times C\frac{\partial^\alpha}{\partial x^\alpha} v_0(x, t) \right) dx \right) dt, \\
&= \int_0^t \left(\int_0^s \left(\delta(x - s) \times C^2\frac{\partial^\alpha}{\partial x^\alpha}\frac{x^{-\alpha-1}}{2^2\Gamma(-\alpha)} \times t \right) dx \right) dt, \\
&+ \int_0^t \left(\int_0^s \left(C\frac{(s - x)^{-\alpha-1}}{2^2\Gamma(-\alpha)} \times t \times C\frac{\partial^\alpha}{\partial x^\alpha} \delta(x) \right) dx \right) dt, \\
&= \left(\frac{C^2}{2^3}\frac{s^{-2\alpha-1}}{\Gamma(-2\alpha)} \times \frac{1}{2}t^2 \right) \\
&+ \int_0^t \left(\int_0^s \left(C\frac{(s - x)^{-\alpha-1}}{2^2\Gamma(-\alpha)} \times t \times C\frac{\partial^\alpha}{\partial x^\alpha} \delta(x) \right) dx \right) dt,
\end{aligned}
$$

$$
\vdots
$$

$$
v_n = \int_0^t \left(\sum_{i=0}^n \sum_{j=0}^n v_i(x, t) * C\frac{\partial^\alpha v_j}{\partial x^\alpha} \right), \qquad i + j = n.
$$

Consequently, the solution of equation (5.8) is given by

$$
u_0(x, t) = v_0(x) = \delta(x),
$$

$$u_1(x,t) = v_0 + v_1 = \delta(x) + C\frac{x^{-\alpha-1}}{2^2\Gamma(-\alpha)} \times t,$$

$$u_2(x,t) = v_0 + v_1 + v_2 = \delta(x) + C\frac{x^{-\alpha-1}}{2^2\Gamma(-\alpha)} \times t + \int_0^t \left(v_0(x,t) * C\frac{\partial^\alpha v_1}{\partial x^\alpha} + v_1(x,t) * C\frac{\partial^\alpha v_0}{\partial x^\alpha}\right) dt,$$

$$\vdots$$

$$u_n(x,t) = v_0 + v_1 + v_2 + \cdots + v_{n+1},$$

$$= \delta(x) + C\frac{x^{-\alpha-1}}{2^2\Gamma(-\alpha)} \times t,$$

$$+ \int_0^t \left(v_0(x,t) * C\frac{\partial^\alpha v_1}{\partial x^\alpha} + v_1(x,t) * C\frac{\partial^\alpha v_0}{\partial x^\alpha}\right) dt + \cdots$$

$$+ \int_0^t \left(\sum_{i=0}^n \sum_{j=0}^n p^{i+j} v_i(x,t) * C\frac{\partial^\alpha v_j}{\partial x^\alpha}\right),$$

where $i + j = n$. So

$$u_n(x,t) = \delta(x) + \sum_{k=1}^n \left(\int_0^t \left(\sum_{i=0}^k \sum_{j=0}^k v_i(x,t) * C\frac{\partial^\alpha v_j}{\partial x^\alpha}\right)\right), \qquad i+j = k.$$

Therefore,

$$u(x,t) = \lim_{n\to\infty} u_n(x,t) = \delta(x) + \sum_{k=1}^\infty \left(\int_0^t \left(\sum_{i=0}^k \sum_{j=0}^k v_i(x,t) * C\frac{\partial^\alpha v_j}{\partial x^\alpha}\right)\right), \qquad i+j = k.$$

$u(x,t)$ is the PDF for the univariate geo-stable distribution.[3] Equation (5.8) gives the special case for the PDF of the geo-stable distribution. The more general fractional PDE for the univariate geo-stable distribution is given by

$$\frac{\partial u}{\partial t} = u(x,t) * \left(-\frac{1+\beta}{2c}\frac{\partial^\alpha}{\partial x^\alpha}u(x,t) - \frac{1-\beta}{2c}\frac{\partial^\alpha}{\partial(-x)^\alpha}u(x,t) + \mu\frac{\partial}{\partial x}u(x,t)\right),$$

$$(5.11)$$

where $0 < \alpha \leq 2$, $\alpha \neq 1$, $-1 \leq \beta \leq 1$, and $-\infty < \mu < \infty$, also $c = \cos\frac{\alpha\pi}{2}$ and $s = \sin\frac{\alpha\pi}{2}$. Given the definition for the HPM, the homotopy for equation (5.11) can be constructed as

$$H(v,p) = (1-p)\left(\frac{\partial v}{\partial t} - \frac{\partial u_0}{\partial t}\right)$$

$$+ p\left(\frac{\partial v}{\partial t} + v(x,t) * d_1\frac{\partial^\alpha v}{\partial x^\alpha} + v(x,t) * d_2\frac{\partial^\alpha v}{\partial(-x)^\alpha} + \mu v(x,t) * \frac{\partial v}{\partial x}\right) = 0,$$

where $d_1 = -\frac{1+\beta}{2c}$ and $d_2 = -\frac{1-\beta}{2c}$.

---

[3] In this case the PDF for the geo-stable distribution is $GS_\alpha((-Ct\cos(\frac{\alpha\pi}{2}))^{\frac{1}{\alpha}}, 1, 0)$.

Suppose the solution of equation (5.11) has the form

$$v = v_0 + v_1 p^1 + v_2 p^2 + v_3 p^3 + \cdots, \qquad (5.12)$$

Substituting (5.12) into (5.11), and comparing coefficients of terms with identical powers of p, leads to

$$p^0 : \frac{\partial v_0}{\partial t} - \frac{\partial u_0}{\partial t} = 0$$

$$p^1 : \frac{\partial v_1}{\partial t} = d_1 v_0(x,t) * \frac{\partial^\alpha v_0}{\partial x^\alpha} + d_2 v_0(x,t) * \frac{\partial^\alpha v_0}{\partial(-x)^\alpha} + \mu v_0(x,t) * \frac{\partial v_0}{\partial x} - \frac{\partial u_0}{\partial t}, \qquad v_1(x,0) = 0$$

$$p^2 : \frac{\partial v_2}{\partial t} = \left(\sum_{i=0}^{1} \sum_{j=0}^{1} v_i(x,t) * d_1 \frac{\partial^\alpha v_j}{\partial x^\alpha} \right) + \left(\sum_{i=0}^{1} \sum_{j=0}^{1} v_i(x,t) * d_2 \frac{\partial^\alpha v_j}{\partial(-x)^\alpha} \right)$$
$$+ \left(\sum_{i=0}^{1} \sum_{j=0}^{1} v_i(x,t) * \mu \frac{\partial v_j}{\partial x} \right), \qquad v_2(x,0) = 0, \quad i+j = 1$$

$$\vdots$$

$$p^{n+1} : \frac{\partial v_{n+1}}{\partial t} = \left(\sum_{i=0}^{n} \sum_{j=0}^{n} v_i(x,t) * d_1 \frac{\partial^\alpha v_j}{\partial x^\alpha} \right) + \left(\sum_{i=0}^{n} \sum_{j=0}^{n} v_i(x,t) * d_1 \frac{\partial^\alpha v_j}{\partial(-x)^\alpha} \right)$$
$$+ \left(\sum_{i=0}^{n} \sum_{j=0}^{1} v_i(x,t) * \mu \frac{\partial v_j}{\partial x} \right), \qquad v_{n+1}(x,0) = 0, \qquad i+j = n$$

For simplicity, we take $v_0(x,t) = u_0(x,t) = \delta(x)$. Consequently, we derive the following recurrence relation

$$v_{n+1} = \int_0^t \left(\left(\sum_{i=0}^{n} \sum_{j=0}^{n} p^{i+j} v_i(x,t) * d_1 \frac{\partial^\alpha v_j}{\partial x^\alpha} \right) \right.$$
$$+ \left(\sum_{i=0}^{n} \sum_{j=0}^{n} v_i(x,t) * d_2 \frac{\partial^\alpha v_j}{\partial(-x)^\alpha} \right)$$
$$\left. + \left(\sum_{i=0}^{n} \sum_{j=0}^{n} v_i(x,t) * \mu \frac{\partial v_j}{\partial x} \right) \right) dt, \qquad i+j = n,$$

for $j = 3, 4, 5, \cdots$. Therefore,

$$u_0(x,t) = v_0(x) = \delta(x),$$

$$u_1(x,t) = v_0 + v_1 = \delta(x) + \int_0^t \left(d_1 v_0(x,t) * \frac{\partial^\alpha v_0}{\partial x^\alpha} + d_2 v_0(x,t) * \frac{\partial^\alpha v_0}{\partial(-x)^\alpha} \right.$$
$$\left. + \mu v_0(x,t) * \frac{\partial v_0}{\partial x} - \frac{\partial u_0}{\partial t} \right) dt,$$

$$u_2(x,t) = v_0 + v_1 + v_2 = \delta(x) + \int_0^t \left(d_1 v_0(x,t) * \frac{\partial^\alpha v_0}{\partial x^\alpha} + d_2 v_0(x,t) * \frac{\partial^\alpha v_0}{\partial(-x)^\alpha} \right.$$

$$\left. + \mu v_0(x,t) * \frac{\partial v_0}{\partial x} - \frac{\partial u_0}{\partial t} \right) dt,$$

$$+ \int_0^t \left(\left(\sum_{i=0}^1 \sum_{j=0}^1 v_i(x,t) * d_1 \frac{\partial^\alpha v_j}{\partial x^\alpha} \right) + \left(\sum_{i=0}^1 \sum_{j=0}^1 v_i(x,t) * d_2 \frac{\partial^\alpha v_j}{\partial(-x)^\alpha} \right) \right.$$

$$\left. + \left(\sum_{i=0}^1 \sum_{j=0}^1 v_i(x,t) * \mu \frac{\partial v_j}{\partial x} \right) \right) dt, i + j = 1$$

$$\vdots$$

$$u_{n+1}(x,t) = v_0 + v_1 + v_2 + \cdots + v_{n+1}$$

$$= \delta(x) + \sum_{k=1}^n \int_0^t \left[\left(\sum_{i=0}^k \sum_{j=0}^k p^{i+j} v_i(x,t) * d_1 \frac{\partial^\alpha v_j}{\partial x^\alpha} \right) + \left(\sum_{i=0}^k \sum_{j=0}^k p^{i+j} v_i(x,t) * d_2 \frac{\partial^\alpha v_j}{\partial(-x)^\alpha} \right) \right.$$

$$\left. + \left(\sum_{i=0}^k \sum_{j=0}^k p^{i+j} v_i(x,t) * \mu \frac{\partial v_j}{\partial x} \right) \right] dt, \qquad i + j = n,$$

and so on. Consequently, the rest of the components of the HPM can be obtained. If $u(x,t) = \lim_{n\to\infty} u_n(x,t)$ and we compute more terms, then we can show that $u(x,t)$ is the PDF of the univariate geo-stable distribution with respect to x, i.e., $GS_\alpha(t^{\frac{1}{\alpha}}, \beta, \mu t)$. Therefore, if $i + j = k - 1$

$$u(x,t) = \lim_{n\to\infty} u_n(x,t)$$

$$= \delta(x) + \sum_{k=1}^\infty \int_0^t \left[\left(\sum_{i=0}^k \sum_{j=0}^k v_i(x,t) * d_1 \frac{\partial^\alpha v_j}{\partial x^\alpha} \right) \right.$$

$$\left. + \left(\sum_{i=0}^k \sum_{j=0}^k v_i(x,t) * d_2 \frac{\partial^\alpha v_j}{\partial(-x)^\alpha} \right) + \left(\sum_{i=0}^k \sum_{j=0}^k v_i(x,t) * \mu \frac{\partial v_j}{\partial x} \right) \right] dt.$$

5.1.2 Adomian Decomposition Method

To solve equation(5.8) with initial condition $u(x,0) = \delta(x)$ by applying the ADM, we construct the following recurrence relation:

$$u_0 = u(x,0) = \delta(x),$$

$$u_{k+1} = \int_0^t \left(\sum_{i=0}^k A_i \right) dt : \qquad k = 0, 1, 2, \cdots,$$

where

$$A_i = \frac{1}{i!}\left[\frac{d^i}{d\lambda^i}\left[\left(\sum_{j=0}^{i}\lambda^i u_i\right)*C\frac{\partial^\alpha}{\partial x^\alpha}\left(\sum_{j=0}^{i}\lambda^i u_i\right)\right]\right]_{\lambda=0} \quad : \quad n=0,1,2,\cdots.$$

Therefore,

$$u_1 = \int_0^t\left(u_0(x,t)*C\frac{\partial^\alpha}{\partial x^\alpha}u_0\right)dt = C\frac{x^{-\alpha-1}}{2^2\Gamma(-\alpha)}\times t,$$

$$u_2 = \int_0^t\left(u_0(x,t)*C\frac{\partial^\alpha u_1}{\partial x^\alpha} + u_1(x,t)*C\frac{\partial^\alpha u_0}{\partial x^\alpha}\right)dt, \qquad u_2(x,0)=0,$$

$$u_3 = \int_0^t\left(\sum_{i=0}^{2}\sum_{j=0}^{2}p^{i+j}u_i(x,t)*C\frac{\partial^\alpha u_j}{\partial x^\alpha}\right)dt, \qquad u_3(x,0)=0, \quad i+j=2, \quad p=1,$$

$$\vdots$$

$$u_{n+1} = \int_0^t\left(\sum_{i=0}^{n}\sum_{j=0}^{n}p^{i+j}v_i(x,t)*C\frac{\partial^\alpha v_j}{\partial x^\alpha}\right)dt, \qquad u_n(x,0)=0, \quad i+j=n, \quad p=1,$$

Consequently,

$$u(x,t) = \lim_{n\to\infty}u_n(x,t) = \delta(x) + \sum_{k=1}^{\infty}\left(\int_0^t\left(\sum_{i=0}^{k}\sum_{j=0}^{k}v_i(x,t)*C\frac{\partial^\alpha v_j}{\partial x^\alpha}\right)\right), \qquad i+j=k-1.$$

An analytic-numerical approximation for the more general class of the univariate geo-stable distribution is obtained by equation (5.11). To solve equation (5.11) with initial condition $u(x,0)=\delta(x)$ using ADM, we construct the following recurrence relation

$$u_0 \quad = \quad u(x,0)=\delta(x),$$

$$u_{k+1} \quad = \quad \int_0^t\left(\sum_{i=0}^{k}A_i\right)dt : \qquad k=0,1,2,\cdots,$$

where

$$A_i = \frac{1}{i!}\left[\frac{d^i}{d\lambda^i}\left[\left(\sum_{j=0}^{i}\lambda^j u_j\right)*d_1\frac{\partial^\alpha}{\partial x^\alpha}\left(\sum_{j=0}^{i}\lambda^j u_j\right) + \left(\sum_{j=0}^{i}\lambda^j u_j\right)*d_2\frac{\partial^\alpha}{\partial(-x)^\alpha}\left(\sum_{j=0}^{i}\lambda^j u_j\right)\right.\right.$$

$$\left.\left.+ \left(\sum_{j=0}^{i}\lambda^j u_j\right)*\mu\frac{\partial}{\partial x}\left(\sum_{j=0}^{i}\lambda^j u_j\right)\right]\right]_{\lambda=0} \quad ,n=0,1,2,\cdots.$$

and $d_1 = -\dfrac{1+\beta}{2c}$ and $d_2 = -\dfrac{1-\beta}{2c}$. Consequently,

$$u_1 = \int_0^t\left(d_1 u_0(x,t)*\frac{\partial^\alpha u_0}{\partial x^\alpha} + d_2 u_0(x,t)*\frac{\partial^\alpha u_0}{\partial(-x)^\alpha} + \mu u_0(x,t)*\frac{\partial u_0}{\partial x} - \frac{\partial u_0}{\partial t}\right)dt, \quad u_1(x,0)=0,$$

$$u_2 = \int_0^t \left(\left(\sum_{i=0}^1 \sum_{j=0}^1 u_i(x,t) * d_1 \frac{\partial^\alpha u_j}{\partial x^\alpha} \right) + \left(\sum_{i=0}^1 \sum_{j=0}^1 u_i(x,t) * d_1 \frac{\partial^\alpha u_j}{\partial(-x)^\alpha} \right) \right.$$

$$\left. + \left(\sum_{i=0}^1 \sum_{j=0}^1 u_i(x,t) * \mu \frac{\partial u_j}{\partial x} \right) \right), \qquad u_2(x,0) = 0, \quad i+j = 1, \quad p = 1,$$

$$\vdots$$

$$u_{n+1} = \int_0^t \left(\left(\sum_{i=0}^1 \sum_{j=0}^1 u_i(x,t) * d_1 \frac{\partial^\alpha u_j}{\partial x^\alpha} \right) + \left(\sum_{i=0}^1 \sum_{j=0}^1 u_i(x,t) * d_1 \frac{\partial^\alpha u_j}{\partial(-x)^\alpha} \right) \right.$$

$$\left. + \left(\sum_{i=0}^1 \sum_{j=0}^1 u_i(x,t) * \mu \frac{\partial u_j}{\partial x} \right) \right), \qquad u_{n+1}(x,0) = 0, \quad i+j = n.$$

Therefore, if $i + j = k - 1$

$$u(x,t) = \lim_{n \to \infty} u_n(x,t)$$

$$= \delta(x) + \sum_{k=1}^\infty \int_0^t \left[\left(\sum_{i=0}^k \sum_{j=0}^k v_i(x,t) * d_1 \frac{\partial^\alpha v_j}{\partial x^\alpha} \right) \right.$$

$$\left. + \left(\sum_{i=0}^k \sum_{j=0}^k v_i(x,t) * d_2 \frac{\partial^\alpha v_j}{\partial(-x)^\alpha} \right) + \left(\sum_{i=0}^k \sum_{j=0}^k v_i(x,t) * \mu \frac{\partial v_j}{\partial x} \right) \right] dt.$$

5.1.3 Variational Iteration Method

To solve equation(5.8) with initial condition $u(x,0) = \delta(x)$, by means of the VIM, we set

$$u_{n+1} = u_n + \int_0^t \lambda \left(\frac{\partial u_n}{\partial s} - v_n(x,t) * C \frac{\partial^\alpha v_n}{\partial x^\alpha} \right) ds, \tag{5.13}$$

so

$$\begin{aligned} \delta u_{n+1} &= \delta u_n + \delta \int_0^t \lambda \left(\frac{\partial u_n}{\partial s} - v_n(x,t) * C \frac{\partial^\alpha v_n}{\partial x^\alpha} \right) ds, \\ &= \delta u_n + \lambda \delta u_n + \int_0^t \left\{ -\frac{d\lambda}{ds} \right\} \delta u_n ds = 0, \end{aligned} \tag{5.14}$$

the stationary conditions of equation (5.14) are:

$$\begin{aligned} 1 + \lambda &= 0, \\ \lambda' &= 0. \end{aligned} \tag{5.15}$$

So the Lagrange multiplier turns out to be as $\lambda = -1$.

Substituting $\lambda = -1$ in (5.13), we get the following variational iteration formula

$$u_{n+1}(x,t) = u_n(x,t) - \int_0^t \left(\frac{\partial u_n}{\partial s} - u_n(x,t) * C \frac{\partial^\alpha u_n}{\partial x^\alpha} \right) ds, \qquad (5.16)$$

where $u_0(x,t) = u(x,0) = \delta(x)$. Consequently,

$$
\begin{aligned}
u_1(x,t) &= u_0(x,t) - \int_0^t \left(\frac{\partial u_n}{\partial s} - u_n(x,t) * C \frac{\partial^\alpha u_n}{\partial x^\alpha} \right) ds, \\
&= \delta(x) + C \frac{x^{-\alpha-1}}{2^2 \Gamma(-\alpha)} \times t,
\end{aligned}
$$

$$
\begin{aligned}
u_2(x,t) &= u_1(x,t) - \int_0^t \left(\frac{\partial u_1}{\partial s} - u_1(x,t) * C \frac{\partial^\alpha u_1}{\partial x^\alpha} \right) ds, \\
&= \delta(x) + \int_0^t \left(v_0(x,t) * C \frac{\partial^\alpha v_1}{\partial x^\alpha} + v_1(x,t) * C \frac{\partial^\alpha v_0}{\partial x^\alpha} \right) dt,
\end{aligned}
$$

where $u_{n+1}(x,t) = u_n(x,t) - v_n(x,t)$ and $u_n(x,0) = 0$, so

$$u_2(x,t) = \delta(x) + \int_0^t \left(v_0(x,t) * C \frac{\partial^\alpha v_1}{\partial x^\alpha} + v_1(x,t) * C \frac{\partial^\alpha v_0}{\partial x^\alpha} \right) dt.$$

$$\vdots$$

Therefore, for $i + j = k - 1$ we have

$$u(x,t) = \lim_{n\to\infty} u_n(x,t) = \delta(x) + \sum_{k=1}^{\infty} \left(\int_0^t \left(\sum_{i=0}^{k-1} \sum_{j=0}^{k-1} v_i(x,t) * C \frac{\partial^\alpha v_j}{\partial x^\alpha} \right) \right).$$

$u(x,t)$ is an approximation for the PDF of the univariate geo-stable distribution. The more general approximation for the PDF of the univariate geo-stable distribution is given by equation (5.11). To solve equation (5.11) by applying the VIM, we set

$$
\begin{aligned}
u_{n+1} = u_n + \int_0^t \lambda \Big(\frac{\partial u_n}{\partial s} &- u_n(x,t) * d_1 \frac{\partial^\alpha u_n}{\partial x^\alpha} \\
&- u_n(x,t) * d_2 \frac{\partial^\alpha u_n}{\partial(-x)^\alpha} - u_n(x,t) * \mu \frac{\partial u_n}{\partial x} \Big) ds,
\end{aligned} \qquad (5.17)
$$

so

$$
\begin{aligned}
\delta u_{n+1} = \delta u_n + \delta \int_0^t \lambda \Big(\frac{\partial u_n}{\partial s} &- v_n(x,t) * d_1 \frac{\partial^\alpha v_n}{\partial x^\alpha} \\
&- v_n(x,t) * d_2 \frac{\partial^\alpha v_n}{\partial(-x)^\alpha} - v_n(x,t) * \mu \frac{\partial v_n}{\partial x} \Big) ds \qquad (5.18) \\
&= \delta u_n + \lambda \delta u_n + \int_0^t \left\{ -\frac{d\lambda}{ds} \right\} \delta u_n ds = 0.
\end{aligned}
$$

The stationary conditions of equation (5.18) are

$$1 + \lambda = 0,$$
$$\lambda' = 0.$$

The Lagrange multiplier turns out to be $\lambda = -1$.

Substituting $\lambda = -1$ in (5.17), we obtain the following variational iteration formula

$$u_{n+1}(x,t) = u_n(x,t) - \int_0^t \left(\frac{\partial u_n}{\partial s} - u_n(x,t) * d_1 \frac{\partial^\alpha u_n}{\partial x^\alpha} \right.$$
$$\left. - u_n(x,t) * d_2 \frac{\partial^\alpha u_n}{\partial (-x)^\alpha} - u_n(x,t) * \mu \frac{\partial u_n}{\partial x} \right) ds,$$

where $u_0(x,t) = u(x,0) = \delta(x)$. Consequently,

$$u_1(x,t) = u_0(x,t) - \int_0^t \left(\frac{\partial u_0}{\partial s} - u_0(x,t) * d_1 \frac{\partial^\alpha u_0}{\partial x^\alpha} \right.$$
$$\left. - u_0(x,t) * d_2 \frac{\partial^\alpha u_0}{\partial (-x)^\alpha} - u_0(x,t) * \mu \frac{\partial u_0}{\partial x} \right) ds,$$

$$u_2(x,t) = \delta(x)$$
$$+ \int_0^t \left(\frac{\partial u_1}{\partial t} - d_1 u_1(x,t) * \frac{\partial^\alpha u_1}{\partial x^\alpha} + d_2 u_1(x,t) * \frac{\partial^\alpha u_1}{\partial (-x)^\alpha} + \mu u_1(x,t) * \frac{\partial u_1}{\partial x} \right) ds$$
$$= \int_0^t \left(\left(\sum_{i=0}^1 \sum_{j=0}^1 v_i(x,t) * d_1 \frac{\partial^\alpha v_j}{\partial x^\alpha} \right) \left(\sum_{i=0}^1 \sum_{j=0}^1 v_i(x,t) * d_1 \frac{\partial^\alpha v_j}{\partial (-x)^\alpha} \right) \right.$$
$$\left. + \left(\sum_{i=0}^1 \sum_{j=0}^1 v_i(x,t) * \mu \frac{\partial v_j}{\partial x} \right) \right) dt, \qquad v_2(x,0) = 0, \qquad i+j = 1.$$

where $u_{n+1}(x,t) = u_n(x,t) - v_n(x,t)$ and $u_n(x,0) = 0$.

$$\vdots$$

As a result, we derive the following recurrence relation

$$v_{n+1} = \int_0^t \left[\left(\sum_{i=0}^n \sum_{j=0}^n p^{i+j} v_i(x,t) * d_1 \frac{\partial^\alpha v_j}{\partial x^\alpha} \right) \right.$$
$$\left. + \left(\sum_{i=0}^n \sum_{j=0}^n v_i(x,t) * d_1 \frac{\partial^\alpha v_j}{\partial (-x)^\alpha} \right) + \left(\sum_{i=0}^n \sum_{j=0}^n v_i(x,t) * \mu \frac{\partial v_j}{\partial x} \right) \right] dt, \qquad i+j = n,$$

for $j = 3, 4, 5, \cdots$.

Consequently, the rest of the components of the VIM can be obtained. If $u(x,t) = \lim_{n\to\infty} u_n(x,t)$ and we compute more terms, then we can show that $u(x,t)$

is the PDF of univariate geo-stable distribution respect to x, i.e., $GS_\alpha(t^{\frac{1}{\alpha}}, \beta, \mu t)$. In other words, for $i + j = k - 1$ we have

$$
\begin{aligned}
u(x, t) &= \lim_{n \to \infty} u_n(x, t) \\
&= \delta(x) + \sum_{k=1}^{\infty} \int_0^t \left[\left(\sum_{i=0}^k \sum_{j=0}^k v_i(x, t) * d_1 \frac{\partial^\alpha v_j}{\partial x^\alpha} \right) \right. \\
&\left. + \left(\sum_{i=0}^k \sum_{j=0}^k v_i(x, t) * d_2 \frac{\partial^\alpha v_j}{\partial (-x)^\alpha} \right) + \left(\sum_{i=0}^k \sum_{j=0}^k v_i(x, t) * \mu \frac{\partial v_j}{\partial x} \right) \right] dt.
\end{aligned}
$$

5.2 Multivariate Geo-stable Distribution

Let $X(t)$ be the position of a particle at time $t > 0$, and $n-$dimensional Euclidean space $\mathbb{R}^n$. Let $P(\mathbf{x}, t)$ denote the density of $X(t)$ where the vector $\mathbf{x} = (x_1, x_2, \cdots, x_n) \in \mathbb{R}^n$. Consider a fractional PDE given by

$$
\frac{\partial P_1(\mathbf{x}, t)}{\partial t} = P_1(\mathbf{x}, t) * (-\langle \mathbf{v}, \nabla P_1(\mathbf{x}, t) \rangle + c \nabla_M^\alpha P_1(\mathbf{x}, t)), \tag{5.19}
$$

where the notation $*$ indicates the convolution operator.

The following theorem explains the connection between multivariate geo-stable distributions and fractional PDEs.

Theorem 5.4   *The fundamental solution of equation (5.19) with the initial condition $P_1(X(0) = 0) = 1$, is the density of the multivariate geo-stable distribution.*

Proof   See Fallahgoul et al. (2014). □

Now, by using the HPM, ADM, and VIM we obtain some analytic-numerical approximations for the PDF of multivariate geo-stable distribution.

5.2.1 Homotopy Perturbation Method

To solve equation (5.19) with initial condition $P_{1_0}(\mathbf{x}, t) = \delta(\mathbf{x})$ using HPM, we construct the following homotopy

$$
H(p, V) = (1 - p) \left(\frac{\partial V}{\partial t} - \frac{\partial P_0}{\partial t} \right) + p \left(\frac{\partial V}{\partial t} + V(\mathbf{x}, t) * (\langle \mathbf{v}, \nabla V \rangle - c \nabla_M^\alpha) \right) = 0. \tag{5.20}
$$

According to equation (5.20), we have

$$
V_0(\mathbf{x}, t) = \delta(\mathbf{x}),
$$

$$
V_1(\mathbf{x}, t) = \int_0^t V_0(\mathbf{x}, t) * (-\langle \mathbf{v}, \nabla V_0 \rangle + c \nabla_M^\alpha V_0) \, dt,
$$

$$
V_2(\mathbf{x}, t) = \int_0^t \sum_{i=0}^1 \sum_{j=0}^1 V_i(\mathbf{x}, t) * (-\langle \mathbf{v}, \nabla V_j \rangle + c \nabla_M^\alpha V_j) \, dt, \qquad i + j = 1,
$$

$$\vdots$$

$$V_{n+1}(\mathbf{x}, t) = \int_0^t \sum_{i=0}^n \sum_{j=0}^n V_i(\mathbf{x}, t) * \left(-\langle \mathbf{v}, \nabla V_j \rangle + c \nabla_M^\alpha V_j \right) dt, \qquad i + j = n.$$

Therefore, the solution is

$$
\begin{aligned}
P_1(\mathbf{x}, t) &= \lim_{p \to 1} V(\mathbf{x}, t), \\
&= V_0(\mathbf{x}, t) + V_1(\mathbf{x}, t) + \cdots + V_{n+1}(\mathbf{x}, t) + \cdots, \\
&= \delta(\mathbf{x}) + \left(\int_0^t V_0(\mathbf{x}, t) * \left(-\langle \mathbf{v}, \nabla V_0 \rangle + c \nabla_M^\alpha V_0 \right) dt \right) + \cdots \\
&+ \int_0^t \sum_{i=0}^n \sum_{j=0}^n p^{i+j} V_i(\mathbf{x}, t) * \left(-\langle \mathbf{v}, \nabla V_j \rangle + c \nabla_M^\alpha V_j \right) dt + \cdots,
\end{aligned}
$$

or

$$P_1(\mathbf{x}, t) = \delta(\mathbf{x}) + \sum_{k=1}^\infty \left[\int_0^t \sum_{i=0}^{k-1} \sum_{j=0}^{k-1} V_i(\mathbf{x}, t) * \left(-\langle \mathbf{v}, \nabla V_j \rangle + c \nabla_M^\alpha V_j \right) dt \right], \, i+j = k-1.$$

5.2.2 Adomian Decomposition Method

Consider the multivariate fractional PDE as equation (5.19). Using ADM to solve that equation with initial condition $P_{1_0}(\mathbf{x}, t) = P_1(\mathbf{x}, 0) = \delta(\mathbf{x})$, we construct the following recurrence relation

$$V_0(\mathbf{x}, t) = \delta(\mathbf{x})$$

$$V_{k+1}(\mathbf{x}, t) = \int_0^t \left(\sum_{i=0}^k A_i \right) dt,$$

$$
\begin{aligned}
A_i = \frac{1}{i!} \Bigg[\frac{d^i}{d\lambda^i} \Bigg[&\left(\sum_{j=0}^i \lambda^i u_i \right) * \langle \mathbf{v}, \nabla \left(\sum_{j=0}^i \lambda^i u_i \right) \rangle \\
&+ c \left(\sum_{j=0}^i \lambda^i u_i \right) * \nabla_M^\alpha \left(\sum_{j=0}^i \lambda^i u_i \right) \Bigg] \Bigg], i = 0, 1, \cdots.
\end{aligned}
$$

According to A_i, we derive the following recurrent relation

$$i = 0 \quad A_0 = V_0(\mathbf{x}, t) * \left(-\langle \mathbf{v}, \nabla V_0 \rangle + c \nabla_M^\alpha V_0 \right),$$

$$i = 1 \quad A_1 = \sum_{i=0}^1 \sum_{j=0}^1 V_i(\mathbf{x}, t) * \left(-\langle \mathbf{v}, \nabla V_j \rangle + c \nabla_M^\alpha V_j \right) dt, \qquad i + j = 1,$$

$$\vdots$$

$$i = n \quad A_n = \sum_{i=0}^n \sum_{j=0}^n V_i(\mathbf{x}, t) * \left(-\langle \mathbf{v}, \nabla V_j \rangle + c \nabla_M^\alpha V_j \right) dt, \qquad i + j = n,$$

Therefore, the solution is

$$
\begin{aligned}
P_1(\mathbf{x}, t) &= V_0(\mathbf{x}, t) + V_1(\mathbf{x}, t) + \cdots + V_{n+1}(\mathbf{x}, t) + \cdots, \\
&= \delta(\mathbf{x}) + \left(\int_0^t V_0(\mathbf{x}, t) * \left(- \langle \mathbf{v}, \nabla V_0 \rangle + c \nabla_M^\alpha V_0 \right) dt \right) + \cdots \\
&\quad + \int_0^t \sum_{i=0}^n \sum_{j=0}^n p^{i+j} V_i(\mathbf{x}, t) * \left(- \langle \mathbf{v}, \nabla V_j \rangle + c \nabla_M^\alpha V_j \right) dt + \cdots,
\end{aligned}
$$

or

$$
P_1(\mathbf{x}, t) = \delta(\mathbf{x}) + \sum_{k=1}^\infty \left[\int_0^t \sum_{i=0}^{k-1} \sum_{j=0}^{k-1} V_i(\mathbf{x}, t) * \left(- \langle \mathbf{v}, \nabla V_j \rangle + c \nabla_M^\alpha V_j \right) dt \right], \quad i+j = k-1.
$$

5.2.3 Variational Iteration Method

Consider the multivariate space-fractional PDE as equation (5.19). To solve that equation with initial condition $P_{1_0}(\mathbf{x}, t) = P_1(\mathbf{x}, 0) = \delta(\mathbf{x})$ utilizing the VIM, we set

$$
V_{n+1}(\mathbf{x}, t) = V_n(\mathbf{x}, t) + \lambda \int_0^t \left(\frac{\partial V_n(\mathbf{x}, s)}{\partial s} + V_n(\mathbf{x}, t) * \left(\langle \mathbf{v}, \nabla V_n(\mathbf{x}, s) \rangle - c \nabla_M^\alpha V_n(\mathbf{x}, s) \right) \right) ds.
$$

In this manner the rest of the components of the VIM can be obtained. If we compute more terms, then we can show that $P_1(\mathbf{x}, t)$ is the PDF of the multivariate geo-stable distribution with respect to $\mathbf{x}$, i.e., $GS_{\alpha, n}(M, \mathbf{v})$. The solution for $i + j = k - 1$ is given by

$$
\begin{aligned}
P_1(\mathbf{x}, t) &= \lim_{n \to \infty} V_n(\mathbf{x}, t), \\
&= V_0(\mathbf{x}, t) + \sum_{k=1}^\infty \left[\int_0^t \sum_{i=0}^{k-1} \sum_{j=0}^{k-1} p^{i+j} V_i(\mathbf{x}, t) * \left(- \langle \mathbf{v}, \nabla V_j \rangle + c \nabla_M^\alpha V_j \right) dt \right], \\
&= \delta(\mathbf{x}) + \sum_{k=1}^\infty \left[\int_0^t \sum_{i=0}^{k-1} \sum_{j=0}^{k-1} V_i(\mathbf{x}, t) * \left(- \langle \mathbf{v}, \nabla V_j \rangle + c \nabla_M^\alpha V_j \right) dt \right].
\end{aligned}
$$

Key points of the chapter

- Geo-stable distributions are useful for modeling heavy tail models.
- Geo-stable distributions do not have a closed-form expression for either their probability density function or cumulative distribution function.
- The probability density function for the geo-stable distributions is the fundamental solution of some fractional diffusion equation for the univariate and multivariate cases.
- The Homotopy perturbation method, Adomian decomposition method, and variational iteration method are useful methods for obtaining the analytic-numerical approximation for some nonlinear problems.

- By applying the Homotopy perturbation method, Adomian decomposition method, and variational iteration method one can obtain good analytic-numerical approximations for the probability density function of the univariate and multivariate stable and geo-stable distributions.

Part II

Applications

6

Fractional Partial Differential Equation and Option Pricing

There are different approaches for studying the behavior of a stochastic process. A stochastic process can be studied as a stochastic differential equation, a partial integro-differential equation, and a fractional partial differential equation. The efficiency of these different approaches depends on the dynamics of the asset price process and the numerical approach for solving them.

Most of the time, solving a stochastic differential equation is not an easy task. However, there are a lot of theories for solving ordinary and partial differential equations. By making a bridge between a stochastic differential equation and ordinary or partial differential equation, one can use the extensive theory of solving ordinary or partial differential equation in solving a stochastic differential equation. By doing this transformation, a stochastic differential equation is equivalent to an ordinary or partial differential equation. There are two ways to doing this transformation: (1) forward Kolmogorov equation, and; (2) backward Kolmogorov equation and the Feynman-Kac formula. The forward Kolmogorov equation is an equation of the probability density of some stochastic process and the backward equation need to be solved backwards in time. It should be noted that the backward equation is an equation for the expected value of some stochastic process.

If the dynamics of the asset price process follows geometric Brownian motion, then the source of randomness is Brownian motion. By assuming the geometric Brownian motion as the source of randomness, Black and Scholes (1973) and Merton (1973) provided a closed-form formula for European call and put options. As we discussed in Chapters 2 and 3, there are several drawbacks of using the Black-Scholes model. Specifically, the Black-Scholes model cannot capture the behavior of the tails and central peak that has been observed for empirical return distributions. Figure (6.1) shows the probability density function for the Gaussian (assumed in the Black-Scholes framework) and a fat-tailed distribution such as the stable distributions. The distributions of a time series for asset returns has been observed to be closer to that of the stable distribution than the Gaussian distribution.

Several alternatives have been suggested for modeling the dynamics of the asset price process. For example, the variance-gamma, normal inverse Gaussian, and CGMY[1] process are useful in capturing the heavy tail property as well as the

---

[1] The CGMY process, also referred to in the literature as the classical tempered stable process, was

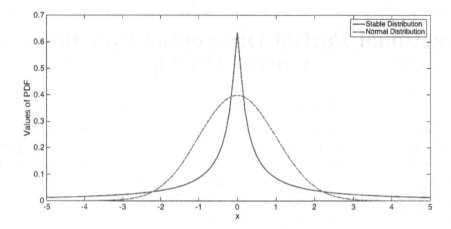

Figure 6.1 The probability density function for the stable and Gaussian distribution.

central peak.[2] It should be noted that all of these process belong to the class of Lévy processes. They are pure-jump processes with infinite activity.[3]

In this chapter, we discuss the details of option pricing within the Black-Scholes framework. Furthermore, the properties of the CGMY process and its advantage in comparison to the geometric Brownian motion assumed in the Black-Scholes framework are explained. Finally, the application of fractional calculus in option pricing by constructing a fractional partial differential equation using the CGMY process is discussed. Bear in mind that those fractional partial differential equations do not have an analytic solution. Instead, numerical approaches such as finite difference and finite element are needed to provide efficient solutions.[4]

In Section 6.1 the Black-Scholes model is discussed. The option pricing problem applying the Lévy process based on a stochastic differential equation, a partial integro-differential equation, and a fractional partial differential equation is then provided in Section 6.2.

---

introduced by Carr et al. (2002). The initials CMGY represent the initials of the four authors (Carr, Geman, Madan, and Yor).

[2] Detailed information about these stochastic models may be found in Samorodnitsky and Taqqu (1994) and Cont and Tankov (2003).

[3] If the number of jumps for a Lévy process is infinite, the process is said to have infinite activity, otherwise it has finite activity. For example, a compound Poisson process is a Lévy process with finite activity and a variance-gamma process is a Lévy process with infinite activity. Bear in mind that a Lévy process with infinite activity can have finite variance.

[4] Detailed information about the application of a numerical solution for the fractional partial differential equation related to option pricing may be found in Cartea and del Castillo-Negrete (2007), Wang et al. (2007), Fang and Oosterlee (2008), Itkin and Carr (2012), and Ludvigsson (2015).

6.1 Option Pricing and Brownian Motion

The dynamics of the asset price process for option pricing based on the Black-Scholes setting is geometric Brownian motion. In this setting, the problem of interest can be formulated either as a stochastic differential equation or a partial differential equation.

6.1.1 Stochastic Differential Equation

A stochastic differential equation is given by

$$X_t = X_0 + \int_0^t \mu(X_u, u)du + \int_0^t \sigma(u, X_u)dB_u \qquad (6.1)$$

where $\mu(X_u, u)$ and $\sigma(u, X_u)$ is stochastic drift and volatility, respectively, and B_t denotes the Brownian motion. The stochastic differential equation given by (6.1) has two parts: a deterministic integral and a stochastic integral. The deterministic part can be solved via Riemann integration theory. However, since the integrand of the stochastic part is a stochastic process, this part cannot be solved using Riemann integration theory. Instead, Itô's lemma for stochastic integration is used for solving the stochastic part.[5]

The stochastic differential equation of the Black-Scholes model for the pricing of a European call option on a non dividend paying stock is given by

$$S_t = S_0 + \int_0^t rS_u du + \int_0^t \sigma S_u dB_u, \qquad (6.2)$$

where S_t, r, and σ are the the stock price, risk-free interest rate, and volatility, respectively. It should be noted that in the Black-Scholes model there are only two types of assets. There is a risk-free asset and a risky asset (a stock).

Two measures for a stochastic process can be considered – physical measure and risk-neutral measure. For option pricing, the risk-neutral measure has to be used. In pricing assets, investors demand a greater potential return for accepting more risk. Therefore, the calculated values for asset prices have to be adjusted for the associated risks. There are two alternatives for making this risk adjustment: (1) taking the expectation under the physical measure and then adjusting the expectation value for the risk, or; (2) adjusting the probabilities of future outcomes by incorporating the effects of risk and then taking the expectation under the adjusted measures. The adjusted probability measure is called risk-neutral measure.

Under a risk-neutral measure, the option price is the expected discounted value of the option's payoff. The risk-neutral measure exists if and only if the market is arbitrage free.[6] Basically, the price process of an option under a risk-neutral measure is given by

$$P(\psi, t) = e^{-r(T-t)} E^{\mathbb{Q}}[\psi], \qquad (6.3)$$

---

[5] See Wiersema (2008).
[6] See Haug (2007).

where T is the option's maturity and ψ is the value function for the option.

The following assumptions are made in the Black-Scholes setting:

- The return distribution is a Gaussian distribution.
- There is no arbitrage in the market.
- Any quantity of the underlying asset and the option can be sold and bought.
- The underlying asset pays no dividends.
- There are no transaction costs.
- The risk-free asset pays a risk-free interest rate.

6.1.2 Partial Differential Equation

Now we turn to the partial differential equation. The link between a partial differential equation and stochastic differential equation (6.2) is the Kolmogorov equation. More specifically, the Kolmogorov backward equation provides a partial differential equation representation for a stochastic differential equation.[7]

Letting $P(\phi, t) = V(S_t, t)$, then by applying the Itô formula to equation $V(S_t, t)$ we have

$$
dV = \frac{\partial V(S_t, t)}{\partial t} dt + \frac{\partial V(S_t, t)}{\partial S_t} dS_t + \frac{1}{2} \frac{\partial^2 V(S_t, t)}{\partial S_t^2} (ds_t)^2,
$$

$$
= \left[\frac{\partial V(S_t, t)}{\partial t} + r S_t \frac{\partial V(S_t, t)}{\partial S_t} + \frac{1}{2} \sigma^2 S_t^2 \frac{\partial^2 V(S_t, t)}{\partial S_t^2} \right] dt + \sigma S_t \frac{\partial V(S_t, t)}{\partial S_t} dB_t,
$$

(6.4)

where $dS_t = \mu S_t dt + \sigma S_t dB_t$ and $S_t^2 = \sigma^2 S_t^2 dt$.

Letting $\phi(t) = \lambda S_t - V(S_t, t)$, then

$$
d\phi = \lambda dS_t + dV(S_t, t),
$$

$$
= \left[\lambda r S_t - \frac{\partial V(S_t, t)}{\partial t} - r S_t \frac{\partial V(S_t, t)}{\partial S_t} - \frac{1}{2} \sigma^2 S_t^2 \frac{\partial^2 V(S_t, t)}{\partial S_t^2} \right] dt
$$

$$
+ \left[\lambda \sigma S_t - \sigma S_t \frac{\partial V(S_t, t)}{\partial S_t} \right] dB_t.
$$

By choosing $\lambda = \dfrac{\partial V(S_t, t)}{\partial S_t}$, we have

$$
d\phi = \left[\lambda r S_t - \frac{\partial V(S_t, t)}{\partial t} - r S_t \frac{\partial V(S_t, t)}{\partial S_t} - \frac{1}{2} \sigma^2 S_t^2 \frac{\partial^2 V(S_t, t)}{\partial S_t^2} \right] dt. \qquad (6.5)
$$

One consequence of equation (6.5) is that the portfolio of risky asset S_t and a

[7] Detailed information about the Kolmogorov equation may be found in Shreve (2004).

risk-free asset paying an interest rate r is riskless, otherwise, there would be an arbitrage opportunity. Therefore, by replacing

$$\phi(t) = \lambda S_t - V(S_t, t),$$

$$\lambda = \frac{\partial V(S_t, t)}{\partial S_t},$$

in equation (6.5), the partial differential equation for a European option[8] is given by

$$\frac{\partial V(S_t, t)}{\partial t} + \frac{1}{2}\sigma^2 S_t^2 \frac{\partial^2 V(S_t, t)}{\partial S_t^2} + r S_t \frac{\partial V(S_t, t)}{\partial S_t} - rV(S_t, t) = 0, \qquad (6.6)$$

with the terminal condition $V(S_t, T) = \phi(S_t)$.

Black and Scholes (1973) and Merton (1973) showed that the following closed-form solutions for equation (6.6) can be obtained for the price of European call and put options

$$C(S_t, t) = G(d_1)S - G(d_2)Ke^{-r(T-t)},$$
$$P(S, t) = Ke^{-r(T-t)} - S + C(S, t),$$
$$= G(-d_2)Ke^{-r(T-t)} - G(-d_1)S,$$

and

$$d_1 = \frac{1}{\sigma\sqrt{T-t}}\left[\ln\left(\frac{S}{K}\right) + \left(r + \frac{\sigma^2}{2}\right)(T-t)\right],$$
$$d_2 = d_1 - \sigma\sqrt{T-t},$$

where $C(S_t, t)$ and $P(S_t, t)$ are respectively the call price and put price, and G, S, and K are the cumulative distribution of the standard Gaussian distribution, spot price, and strike price, respectively.

The Feynman-Kac theorem provides tools for obtaining the solution to a partial differential equation as a conditional expectation with respect to a stochastic process. In other words, an option can be priced given a payoff function of time by finding the solution to a differential equation without concern for the stochastic process. The Feynman-Kac theorem gives coefficients for the first-order partial derivative $V(S_t, t)$ with respect to t and the second-order derivative $V(S_t, t)$ with respect to S_t.

Now, we discuss the details of the Feynman-Kac theorem for the Black-Scholes model. In fact, a stochastic differential equation (6.2) is obtained from a partial differential equation (6.4).

If the discounting value for the option value $V(S_t, t)$ is

$$Z_t = V(S_t, t)e^{-qt},$$

---

[8] A European option is one in which the option buyer can only exercise the option at the contract's expiration date.

then

$$\frac{\partial Z}{\partial t} = -qV(S_t, t)e^{-qt}, \quad \frac{\partial Z}{\partial V} = e^{-qt}, \quad \frac{\partial^2 Z}{\partial V^2} = 0, \quad \text{and} \quad dt dV = 0.$$

By applying the Itô lemma to Z_t, dZ_t is then

$$dZ_t = \frac{\partial Z_t}{\partial t}dt + \frac{\partial Z_t}{\partial V}dV + \frac{1}{2}\frac{\partial^2 Z_t}{\partial V^2}(dV)^2 + \frac{\partial^2 Z_t}{\partial t \partial V}dt dV,$$

$$= e^{-qt}\left[-rV(S_t, t) + \frac{\partial V(S_t, t)}{\partial t} + rS_t\frac{\partial V(S_t, t)}{\partial S_t} + \frac{1}{2}\sigma^2 S_t^2\frac{\partial^2 V(S_t, t)}{\partial S_t^2}\right]dt$$

$$+ e^{-qt}\sigma S_t\frac{\partial V(S_t, t)}{\partial S_t}dB_t. \tag{6.7}$$

If the drift of $V(S_t, t)$ is 0, then the integral form of Z_t over interval $[0, T]$ is given by

$$Z_T = Z_0 + \int_0^T e^{-qt}\sigma S_t\frac{\partial V(S_t, t)}{\partial S_t}dB_t. \tag{6.8}$$

The expectation of equation (6.8) is equal to

$$E\Big[Z_T\Big] = Z_0 + E\left[\int_0^T e^{-qt}\sigma S_t\frac{\partial V(S_t, t)}{\partial S_t}dB_t\right].$$

Therefore,

$$V(S_t, 0) = E\Big[e^{-qT}V(S_t, T)\Big] \tag{6.9}$$

One interpretation of equation (6.9) is that the initial value of the option can be determined as the expected value of the discounted random terminal value. It should be noted that we used the assumption that the drift part of equation (6.7) is zero. In other words,

$$\left[-rV(S_t, t) + \frac{\partial V(S_t, t)}{\partial t} + rS_t\frac{\partial V(S_t, t)}{\partial S_t} + \frac{1}{2}\sigma^2 S_t^2\frac{\partial^2 V(S_t, t)}{\partial S_t^2}\right]dt = 0,$$

This is equivalent to equation (6.6). Therefore, the dynamics of the asset price process, in the expected value expression, has r as the drift coefficient and it is given by

$$dS_t = rS_t dt + \sigma S_t dB_t.$$

Volatility plays a critical role in asset pricing.[9] There is a one-to-one relation between volatility, $\sigma(S_t, t)$, and the option price. In fact, by observing the option's market price, one can compute the value of volatility. A volatility that is obtained from the option market price is called implied volatility.

The implied volatility curves for the Black-scholes model is flat. This is in stark

_____
[9] See, among others, Javaheri (2011).

contrast to empirical data based on option prices that clearly indicate that the implied volatility curve is not flat.

Equation (6.3) shows that the option price depends on the risk-neutral measure. More specifically, because of $E^{\mathbb{Q}}[\phi]$, the option price is obtained by the risk-neutral probability density function.

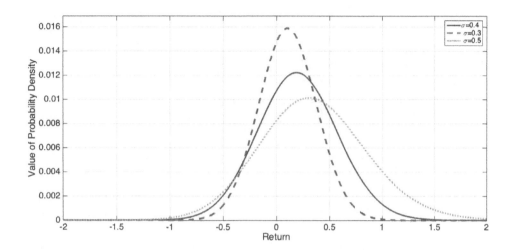

Figure 6.2 Different risk-neutral probability density function for the Black-Scholes model.

Figure 6.2 shows the risk-neutral probability density function for the Black-Scholes model. As mentioned at the outset of this chapter, these probability density functions do not describe precisely the properties of tails as well as the central peak of observed asset return distributions. It should be noted that these probability density functions are the Gaussian's probability density functions.

6.2 Option Pricing and the Lévy Process

Several stochastic processes are proposed as alternatives to geometric Brownian motion for option pricing. Providing a rich class of stochastic processes, the exponential Lévy process describes well the stylized facts that have been observed regarding the distribution of asset prices. That is, the exponential Lévy process is characterized by heavy tails and excess kurtosis.[10] The variance-gamma, normal inverse Gaussian, stable, log-stable, and CGMY processes are important examples of the exponential Lévy process. These processes are pure-jump processes with infinite activity. It should be noted that the Brownian motion is an example of a Lévy process.[11] In this section, we discuss the application of the CGMY process for option pricing.

[10] Excess kurtosis is kurtosis larger than 3.
[11] Detailed information about these classes of stochastic process can be found in Cont and Tankov (2003).

6.2.1 Lévy Process

The CGMY process, suggested by Carr et al. (2002), is a Lévy process. Before discussing the CGMY process, we first provide the definition and some properties of the Lévy process.

Definition 6.1 Let $(X_t)_{t \geq 0}$ be a stochastic process on a probability space. X_t is a Lévy process if it has the following properties

- $X_0 = 0$.
- X_t has independent increments.
- X_t has stationary increments.
- X_t is stochastically continuous. More precisely, $\forall t \geq 0$ and $a > 0$

$$\lim_{s \to t} P\left[|X_s - X_t| > a\right] = 0.$$

- X_t is a cadlag.

A Lévy process is fully determined by a triple, referred to as the Lévy triple. The Lévy triple is denoted by the triple (σ^2, ν, γ), where the parameter σ and γ are real numbers and refer to the variation and location of the process, respectively. The parameter ν is a special measure that is referred to as the Lévy measure.

The closed-form formula for almost all classes of the Lévy process do not exist. Instead, the characteristic function for the Lévy process can be obtained in closed-form. The characteristic function plays a major role in the empirical application of the Lévy process. The characteristic function of the univariate Lévy process X_t, obtained by using the Lévy-Khintchine formula,[12] is given by

$$\Phi(u; X_t) = e^{\phi(u, X_t)t},$$

where

$$\phi(u, X_t) = \left(i\gamma u - \frac{1}{2}\sigma^2 u^2 + \int_{-\infty}^{\infty} \left(e^{iux} - 1 - iux\mathbf{1}_{|x| \leq 1} \right) \nu(dx) \right), \qquad (6.10)$$

where $\mathbf{1}_{|X|}$ is an indicator function.[13] $\phi(u, X_t)$ is called the characteristic exponent.

If $\nu(dx) = 0$, the characteristic function is equal to the characteristic function of Brownian motion. Moreover, if $\sigma = 0$ the process is referred to as a pure-jump process. Therefore, the exponential exponent of a pure-jump process, X_t, is[14]

$$\phi(u; X_t) = \exp\left(i\gamma u + \int_{-\infty}^{\infty} \left(e^{iux} - 1 - iux\mathbf{1}_{|x| \leq 1} \right) \nu(dx) \right).$$

---

[12] The Lévy-Khintchine formula is a useful tool for obtaining the characteristic function of all types of Lévy processes. More information can be found in Cont and Tankov (2003).

[13] When $\mathbf{1}_B$ is an indicator function, then for any set as B, $\mathbf{1}_B = 1$ for $x \in B$ and $\mathbf{1}_B = 0$ for $x \neq B$.

[14] It should be noted that the Poisson, gamma, stable, and all classes of tempered stable process are a pure-jump process.

6.2.2 CGMY Process

The Lévy measure for the CGMY process is given by

$$\nu(dx) = \left(C\frac{e^{-Mx}}{x^{1+Y}}\mathbf{1}_{x>0} + C\frac{e^{-G|x|}}{|x|^{1+Y}}\mathbf{1}_{x<0} \right), \tag{6.11}$$

where $C, G, M > 0, 0 \leq Y < 2$. These parameters fully describe the properties of the CGMY process. In other words, G and M show the rate of decay of the left and right tail, respectively. Y is the exponential exponent which plays a major role in determining the rate of decay of the tails.

The CGMY processes are obtained by tempering the tails of stable processes. For this reason, it is also referred to as the classical tempered stable process. There is a relationship between the Lévy measure for the stable process and CGMY process. By setting $G = M = 0$ and choosing different values for parameter C for the left and right one can achieve the Lévy measure of the stable process. A drawback of the stable process is that only the first moment exists. By tempering the tails of the stable process, a finite moment of any order can be obtained. Some researchers use the tempered stable process instead of the CGMY process. Detailed information about the stable and tempered stable process can be found in Samorodnitsky and Taqqu (1994) and Rachev et al. (2011).

Figure 6.3 Different trajectories for the CGMY process.

6.2.3 Stochastic Differential Equation

If the dynamics of the asset price process are characterized by the semimartingale property, the stochastic differential equation of the option price represents a useful model for empirical application. More specifically, the absence of arbitrage is equivalent with semimartingale property. The stochastic differential equation for the option price when the dynamics of the asset price is a CGMY process can be obtained by constructing a simple portfolio with two risky and risk-free assets. More precisely, let N_t be the price process for the risk-free asset and S_t the log-price process of a risky asset, then the stochastic differential equation for the CGMY process is given by

$$dN_t = rN_tdt,$$
$$dS_t = (r - d + \omega_S)dt + dX_t, \tag{6.12}$$

where X_t, r, and d are the CGMY process, risk-free interest rate, and dividend, respectively, and

$$\omega_S = \Phi(-i, X_t).$$

where ω_t is the martingale correction. In other words, by choosing a suitable value for ω one can be sure that the dynamics of the asset price is a martingale under the risk-neutral measure.[15] It should be noted that the first fundamental theorem of asset pricing shows the necessary and sufficient conditions for absence of arbitrage in a market. It shows that there exists a risk-neutral measure, which is equivalent to the physical measure.[16]

Since the closed-form formula for the risk-neutral density does not exist, one cannot apply equation (6.3). There are two approaches that can be employed to overcome this problem: Monte Carlo simulation and fast Fourier transform.

A considerable number of researchers and practitioners are using the fast Fourier transform for option pricing. It is an efficient approach. However, Monte Carlo simulation is an easy and straightforward method. Although this method utilizes the cumulative distribution function, the closed-form formula for the cumulative distribution function for the CGMY process does not exist. Instead, the closed-form formula for the characteristic function can be obtained by using Lévy-Khinchin formula. There are some numerical approaches for calculating the cumulative distribution function based on its characteristic function.[17]

The rest of this section describes how using the characteristic function, the probability density function, and the cumulative distribution function can be calculated.

Let X_t be a stochastic process, then its characteristic function, $E[e^{iuX_t}]$, is the Fourier transform of its probability density function. Consequently, by obtaining the inverse of the related Fourier transform one can obtain the probability density function. This approach was introduced by DuMouchel (1975) for calculating the probability density function of the stable distribution. Later this approach was applied to some classes of the tempered stable distribution in Kim et al. (2009).

---

[15] Some researchers are using risk-neutral dynamics instead of the dynamics of asset prices. In this chapter, when we refer to the dynamics of the asset price we mean risk-neutral dynamics.

[16] Detailed information about the risk-neutral and physical measure can be found in Cont and Tankov (2003).

[17] Detailed information about these methods can be found in Haug (2007) and Cont (2005).

Probability Density Function

Let X be a random variable of a stochastic process, then its characteristic function is given by

$$\Phi(u; X) = E[e^{iuX}] = \int_{-\infty}^{\infty} e^{iux} f_X(x) dx,$$

where $f_X(X)$ is the probability density function for the random variable X. Consequently, the probability density function for random variable X is given by

$$f_X(x) = \int_{-\infty}^{\infty} e^{-iux} \Phi(u; X) du, \tag{6.13}$$

where $\Phi(u; X)$ is the characteristic function. Therefore, by calculating equation (6.13), one can obtain the probability density function for random variable X. Since the characteristic function for the CGMY process is complicated, an analytic formula for equation (6.13) does not exist. However, useful numerical approaches are suggested based on numerical integration and fast Fourier transform.[18] The first step in numerical approximation for equation (6.13) is the discrete Fourier transform, which transfers vector $Y = (y_1, y_2, \cdots, y_n)$ to vector $X = (x_1, x_2, \cdots, x_n)$. More specifically, let $Y = (y_1, y_2, \cdots, y_n), X = (x_1, x_2, \cdots, x_n) \in \mathbb{R}^n$, then the transformation is given by

$$x_j = \sum_{k=1}^{n} y_i e^{\frac{2\pi(j-1)(k+1)}{n}}, \quad j = 1, 2, \cdots, n. \tag{6.14}$$

The computational cost of equation (6.14) is high, even using a high-speed computer. Instead, the fast Fourier transform provides a convenient way to overcome the speed problem when employing the discrete Fourier transform.

Let $a \in \mathbb{R}^+, q \in \mathbb{N}^+$ and for $j, k \in \{1, 2, \cdots, n = 2^q\}$ given the following assumptions

$$u_k = -a + \frac{2a}{n}(k - 1),$$

$$u_k^* = \frac{u_{k+1} - u_k}{2},$$

$$x_j = -\frac{n\pi}{2a} + \frac{\pi}{a}(j - 1),$$

$$C_j = \frac{a}{n\pi}(-1)^{j-1} e^{i\frac{\pi(j-1)}{n}}.$$

An approximation for equation (6.13) at x_j based on the fast Fourier transform

---

[18] See, among others, Bailey and Swarztrauber (1994).

is given by

$$f_X(x_j) = \frac{1}{2\pi} \int_{-\infty}^{\infty} e^{-iux_j} \Phi(u; X) du,$$

$$\sim \frac{1}{2\pi} \int_{-a}^{a} e^{-iux_j} \Phi(u; X) du,$$

$$\sim C_j \sum_{i=1}^{n} (-1)^{k-1} \Phi(u_k^*; X) e^{-i\frac{2\pi(j-1)(k-1)}{n}},$$

where parameters a and q are the limit for integration of the Fourier transform and defines the number of integration steps, respectively.[19]

Cumulative Distribution Function

Evaluating the cumulative distribution function is almost similar to calculating the probability density function. The difference between computing the cumulative distribution function and the probability density function is related to the form of the characteristic function. For example, Kim et al. (2010) show that the cumulative distribution function for the tempered stable distribution is given by

$$F_X(x) = \frac{e^{x\rho}}{\pi} \Re \left(\int_0^{\infty} e^{-iux} \frac{\Phi(u + i\rho; X)}{c(\rho) - iu} du \right), \qquad x \in \mathbb{R}$$

where for all complex numbers Z, $|\phi_X(Z)| < \infty$ and $\Im(Z) = \rho > 0$. Bear in mind that $\Re$ and $\Im$ are the real and imaginary part of a complex number, respectively. For simplicity, let $g_X(u) = \dfrac{\Phi(u + i\rho; X)}{\rho - iu}$. In order to apply the discrete and fast Fourier transform, let $a \in \mathbb{R}^+, q \in \mathbb{N}^+$ and for $j, k \in \{1, 2, \cdots, n = 2^q\}$ given the following assumptions

$$u_k = \frac{2a}{n}(k - 1),$$

$$u_k^* = \frac{u_k + u_{k+1}}{2} = \frac{a}{n}(2k - 1),$$

$$x_j = -\frac{n\pi}{2a} + \frac{\pi}{a}(j - 1).$$

Therefore,

$$\int_0^{\infty} e^{-ix_j u} g_X(u) du \sim \int_0^{a} e^{-ix_j u} g_X(u) du,$$

$$\sim \sum_{k=1}^{n} e^{-ix_j u_k^*} g_X(u_k^*) \frac{2a}{n},$$

---

[19] More details about the accuracy of this approach for the stable distribution can be found in Menn and Rachev (2006).

and by simple computation, one can show for each x_j, $F_X(x_j)$ is equal to

$$F_X(x_j) \sim \frac{e^{x_j \rho}}{\pi} \Re \left\{ \frac{2a}{n} e^{-i\frac{\pi}{n}(j-1)} \sum_{k=1}^{n} e^{i\pi\frac{(2k-1)}{2}} g_X(u_k^*) e^{-i\frac{2\pi}{n}(k-1)(j-1)} \right\}.$$

It should be noted that the accuracy of this approximation depends on the numerical integration method that is used.[20]

Figure 6.3 shows different trajectories for the CGMY process for 10 years of simulated log prices. It reveals that the CGMY process is a pure-jump process. The implied volatility curve for the CGMY process is more realistic because volatility exhibits the well-known smile and skewness observed in the market.[21] Figure 6.4 shows the implied volatility curves for the CGMY process using the simulated data from CGMY process.[22] By comparing the implied volatility curves of Black-Scholes, one can see that the implied volatility curves for the CGMY process are not flat and they have the volatility smile property.

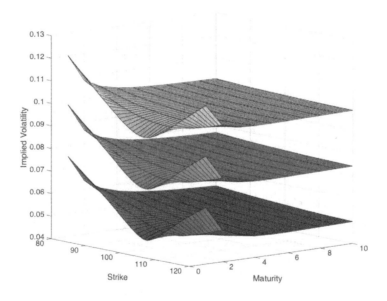

Figure 6.4 Implied volatility curves for the CGMY process.

Figure 6.5 shows the simulated risk-neutral probability density function for the

[20] The accuracy and speed of this approximation can be improved by using different numerical integration methods. More detailed information can be found in Menn and Rachev (2006) and Kim et al. (2010).

[21] More information about the smile and skewness of the implied volatility curve can be found in Javaheri (2011).

[22] For obtaining the implied volatility curve for the CGMY model, one of the following approaches should be used: stochastic differential equation, partial integro-differential equation, and fractional partial differential equation.

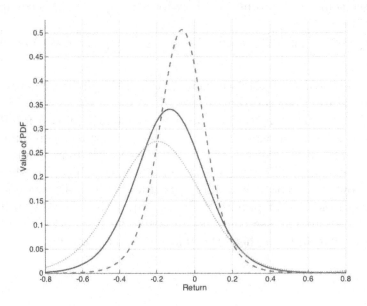

Figure 6.5 Risk-neutral probability density function for the simulated CGMY process.

simulated CGMY process. It reveals that the risk-neutral density has heavy tails and excess kurtosis. In fact, the CGMY process overcomes some drawbacks of geometric Brownian motion assumed in the Black-Scholes setting.

6.2.4 Partial Integro-Differential Equation

By replacing the dynamics of the asset price process from geometric Brownian motion to an exponential Lévy process – variance-gamma, normal inverse Gaussian, and CGMY process – some drawbacks of the Black-Scholes model can be resolved. However, because of infinite activity and the existence of jumps, a closed-form formula for the price of different types of options does not exist.

One approach for solving the stochastic differential equation given by equation (6.2) is using the Feyman-Kac theorem. More precisely, by the change of variable method,[23] a partial differential equation equivalent to the stochastic differential equation (6.2) can be achieved. However, if the dynamics of the asset price process follows an exponential Lévy process, by using the change of variable method the equivalent problem of the related stochastic differential equation is a partial integro-differential equation.[24]

In this section, we first discuss the details of constructing a partial integro-

---

[23] It is basically the reverse of the chain rule.
[24] Detailed information about these partial integro-differential equations can be found in Cont and Tankov (2003).

differential equation from the stochastic differential equation (6.2), where the dynamics of the asset price process is an exponential Lévy process, and then we move on to solve the related partial integro-differential equation. Before introducing the partial integro-differential equation for the exponential Lévy process, the definition of a compensated process is given.

Definition 6.2 Let X_t be a stochastic process and ω_t be a deterministic value such that $\tilde{X}_t = X_t - \omega_t$ is a martingale. Then $\tilde{X}_t$ is called a compensated process and the deterministic part ω_t is said to be a compensator.

Let S_t and $\tilde{S}_t$ be the dynamics of the asset price process and a compensated process of S_t, respectively. Then the stochastic differential equation (6.12) is given by

$$S_t = S_0 + \int_0^t (r-d)S_u du + \int_0^t S_{u-}\sigma dB_u + \int_0^t \int_{-\infty}^\infty (e^x - 1)S_{u-}\tilde{J}_X(du, dx),$$

where $\tilde{J}_X(du, dx)$ denotes the compensated jump measure for the Lévy process X_t. Equivalently, for the compensated process $\tilde{S}_t$, the following stochastic differential equation can be obtained

$$\tilde{S}_t = \tilde{S}_0 + \int_0^t \tilde{S}_{u-}\sigma dB_u + \int_0^t \int_{-\infty}^\infty (e^x - 1)\tilde{S}_{u-}\tilde{J}_X(du, dx). \tag{6.15}$$

Based on equation (6.3), the value of a European option with the Lévy process risk-neutral density is

$$P(S, t) = e^{-(r-d)(T-t)} E^{\mathbb{Q}}[\psi(S_T)|\mathcal{F}_t],$$
$$= e^{-(r-d)(T-t)} E^{\mathbb{Q}}[\psi(S_T)|S_t = S],$$

where $\psi(S_T)$ is the terminal payoff.

Now, by the change of variable method, the following equivalent problem for equation (6.15) can be obtained

$$\tau = T - t,$$
$$x = ln\left(\frac{S}{K}\right) + r\tau,$$
$$h(x) = \frac{\psi(Ke^x)}{K},$$
$$u(\tau, x) = \frac{e^{(r-d)(T-t)}P(S, t)}{K},$$

then

$$u(\tau, x) = E^{\mathbb{Q}}[h(x + X_\tau)].$$

Therefore, by assuming that h is in the domain of the infinitesimal generator,[25]

---

[25] Detailed information about the infinitesimal generator can be found in Cont and Tankov (2003).

we have

$$\frac{\partial P(S,t)}{\partial t} + rS\frac{\partial P(S,t)}{\partial S} + \frac{\sigma^2 S^2}{2}\frac{\partial^2 P(S,t)}{\partial S^2} - rP(S,t)$$

$$+ \int_{-\infty}^{\infty}\left(P(Se^y,t) - P(S,t) - S(e^y - 1)\frac{\partial P(S,t)}{\partial S}\right)\nu(dy),$$

subject to terminal condition $P(x,T) = h(x)$, where $\nu(dy)$ is the related Lévy measure.[26]

European Option

A European option with terminal payoff $\psi(S_t)$, is $P(S,t)$ and it is the solution of

$$\frac{\partial P(S,t)}{\partial t} + rS\frac{\partial P(S,t)}{\partial S} + \frac{\sigma^2 S^2}{2}\frac{\partial^2 P(S,t)}{\partial S^2} - rP(S,t)$$

$$+ \int_{-\infty}^{\infty}\left(P(Se^y,t) - P(S,t) - S(e^y - 1)\frac{\partial P(S,t)}{\partial S}\right)\nu(dy),$$

with the terminal condition

$$\forall S > 0 \quad P(S,T) = \psi(S).$$

Letting

$$f(y,t) = e^{(r-d)(T-t)}P(S_0 e^y),$$

$$y = log\left(\frac{S}{S_0}\right),$$

then the forward value of option f is the solution to the following partial integro-differential equation

$$\frac{\partial f(y,t)}{\partial t} + \left(r - \frac{\sigma^2}{2}\right)\frac{\partial f(y,t)}{\partial y} + \frac{\sigma^2}{2}\frac{\partial^2 f(y,t)}{\partial y^2}$$

$$+ \int_{-\infty}^{\infty}\left(f(y+z,t) - f(y,t) - S(e^z - 1)\frac{\partial f(y,t)}{\partial y}\right)\nu(dz),$$

subject to

$$\forall x \in \mathbb{R} \quad f(x,T) = h(x).$$

Since the above partial integro-differential equations do not have an analytic solution, numerical methods such as finite difference, finite element, and spectral method are used.[27]

---

[26] Detailed information about constructing the partial integro-differential equation for different types of option can be found in Chapter 12 in Cont and Tankov (2003).

[27] Here we do not discuss the application of the finite difference and finite element methods. Detailed information about these methods can be found, in Cont and Tankov (2003) and Schwab et al. (2007).

6.2.5 Fractional Partial Differential Equation

In this section, we describe the derivation of the fractional partial differential equation for three classes of Lévy process: log-stable process, CGMY, and KoBoL process.[28] These fractional partial differential equations can be obtained either using a characteristic function or the change of variable method.[29] Since these fractional partial differential equations do not have an analytic solution, numerical approaches such as finite difference and finite element are used by practitioners.[30] Bear in mind that since the closed-form formula for the characteristic function for those processes are available, there are alternative methods such as fast Fourier transform and Monte Carlo simulation.

The most popular approach for approximating solutions for the fractional partial differential equation is the finite difference method. However, some classes of fractional partial differential equations that arise in option pricing are more complicated and, as a result, the finite difference method does not give an accurate approximation. In this case, splitting the integral operator has been suggested.[31] More specifically, Itkin (2013) transformed the jump integral into a pseudo-differential operator. Then for various jump models, Itkin shows how to construct an appropriate first- and second-order approximation on a grid.

Let S_t be the price process, $X_t = \log(S_t)$, and the dynamics of the asset price process be a Lévy process. Then the stochastic differential equation for the corresponding option pricing model is given by

$$dN_t = rN_t dt,$$
$$S_t = S_0 e^{(r-d)t - \omega_S t + L_t}.$$

For simplicity we set $d = 0$. The price process for period $[t, T]$ is then

$$S_T = S_t + e^{(r-\omega_S)(T-t)+L_t},$$
$$\log(S_T) = \log(S_t) + (r - \omega_S)(T - t) + L_t,$$
$$\log(S_T) - \log(S_t) = e^{(r-\omega_S)(T-t)+L_t},$$
$$X_T - X_t = (r - \omega_S)(T - t) + \int_t^T dL_t,$$
$$dX_t = (r - \omega_S)dt + dL_t. \tag{6.16}$$

The fractional partial differential equation for a European and an American option can be constructed by using the characteristic exponent. In the next section, the details of the characteristic function for the log-stable, CGMY, and

---

[28] KoBoL process belongs to the class of tempered stable processes. More precisely, by setting $M = G = \lambda$ for the CGMY process, the KoBoL process can be obtained. Detailed information about this process may be found in Boyarchenko and Levendorski (2000).

[29] Cartea and del Castillo-Negrete (2007) use the characteristic function for constructing the fractional partial differential equation.

[30] Detailed information about the application of numerical approaches for constructing a fractional partial differential equation can be found in Meerschaert and Tadjeran (2004), Meerschaert and Tadjeran (2006), and Andersen and Lipton (2013).

[31] See, among others, Carr and Mayo (2007), Itkin and Carr (2012), and Itkin (2013).

KoBoL processes are given. It should be noted that although we only discuss the details of the fractional partial differential equation for these processes, by applying the same approach the fractional partial differential equation for the other Lévy processes such as the variance-gamma process can be obtained.

Characteristic Function

The characteristic function of a random variable X is the Fourier transform of the probability density function of a random variable X. More precisely, let X be a random variable with probability density function $f_X(x)$. Then its characteristic function is given by

$$\Phi(u; X) = \int_{-\infty}^{\infty} e^{iux} f_X(x) dx.$$

In order to obtain the characteristic function for a Lévy process, the Lévy-Khintchine formula, equation (6.10), plays an important role. More precisely, by putting the Lévy measure into equation (6.10), the characteristic function can be calculated.

The Lévy measure for the log-stable process, is given by

$$\nu_{LS}(dx) = \left(C \frac{p}{x^{1+Y}} \mathbf{1}_{x>0} + C \frac{q}{|x|^{1+Y}} \mathbf{1}_{x<0} \right),$$

where $C > 0$, $p, q \in [-1, 1]$ and $p + q = 1$. By using the Lévy-Khintchine formula, the characteristic function for the log-stable is given by

$$\Phi(u; LS) = \exp\left(\phi(u; LoS_t) t \right),$$

$$= \exp\left(\left\{ \frac{\sigma^\alpha}{4\cos(\alpha\pi/2)} \left[(1 - \beta)(iu)^\alpha + (1 + \beta)(-iu)^\alpha \right] + i\gamma u \right\} t \right).$$

It should be noted that the characteristic function depends on the values for α, β, and σ.[32] By setting $\beta = -1$, the characteristic exponent of the log-stable process is

$$\phi(-u; LS) = -\frac{1}{2}\sigma^\alpha \sec\left(\frac{\alpha\pi}{2} \right) (-iu)^\alpha. \tag{6.17}$$

By using equations (6.11) and (6.10), the characteristic function for the CGMY process is given by

$$\Phi(u; CGMY) = \exp\left(\phi\left(u; CGMY\right) t \right),$$

$$= \exp\left(C\Gamma(Y)\left((M - iu)^Y - M^Y + (G + iu)^Y - G^Y \right) t \right),$$

where $\Gamma(.)$ is the gamma function.

---

[32] Detailed information about the characteristic function for the log-stable process can be found in Carr and Wu (2003).

The Lévy measure for the KoBoL process is

$$\nu_{KoBoL}(dx) = \left(C \frac{pe^{-\lambda x}}{x^{1+Y}} \mathbf{1}_{x>0} + C \frac{qe^{-\lambda|x|}}{|x|^{1+Y}} \mathbf{1}_{x<0} \right),$$

and therefore its characteristic function is given by

$$\Phi(u; KoBoL) = \exp\left(\phi\left(u; KoBoL\right)t \right),$$

$$= \exp\left(\frac{1}{2}\sigma^\alpha \left(p(\lambda - iu)^\alpha + q(\lambda + iu)^\alpha - \lambda^\alpha \right) \right), \quad \text{for} \quad 0 < \alpha < 1$$

When $1 < \alpha \le 2$,

$$\Phi(u; KoBoL) = \exp\left(\phi\left(u; KoBoL\right)t \right),$$

$$= \exp\left(\frac{1}{2}\sigma^\alpha \left(p(\lambda - iu)^\alpha + q(\lambda + iu)^\alpha - \lambda^\alpha - iu\alpha\lambda^{\alpha-1}(q - p) \right)t \right).$$

From structure of the Lévy measure for the CGMY and KoBoL processes it is revealed that their Lévy measures are obtained by tempering the tails of the stable process. The process obtained from the stable process by tempering the tail is called the tempered stable process. In fact, by using different tempered functions, one can define different tempered stable processes. Detailed information about the Lévy measure and characteristic function for the tempered stable process can be found in Rachev et al. (2011) and Fallahgoul et al. (2016).

European Option

The fractional partial differential equation for the stochastic differential equation (6.16) is obtained by inverting the Fourier transform of the value of the option. More specifically, if $P(S, t)$ denotes the value of a call or put European option, then the value is given by

$$P(S, t) = e^{-(r-\omega_S)(T-t)} E^{\mathbb{Q}} \left[\psi(S_T) | S_t = S \right].$$

where $\psi(S_T)$ is the option's terminal payoff. The Fourier transform of the terminal payoff is

$$\Phi\left(u; \psi(S_T)\right) = \int_{-\infty+iu_i}^{\infty+iu_i} e^{iuS_T} \psi(S_T) dS_T,$$

and therefore,

$$P(S,t) = \frac{e^{-(r-\omega_S)(T-t)}}{2\pi} E^{\mathbb{Q}} \left[\int_{-\infty+iu_i}^{\infty+iu_i} e^{-iuS_T} \Phi\left(u; \psi(S_T)\right) du \Big| S_t = S \right],$$

$$= \frac{e^{-(r-\omega_S)(T-t)}}{2\pi} \int_{-\infty+iu_i}^{\infty+iu_i} E^{\mathbb{Q}} \left[e^{-iuS_T} \Big| S_t = S \right] \Phi\left(u; \psi(S_T)\right) du,$$

$$= \frac{e^{-(r-\omega_S)(T-t)}}{2\pi} \int_{-\infty+iu_i}^{\infty+iu_i} e^{-iuS_T - iu(r-\omega_S)(T_t)} e^{(T-t)\phi(u;L_t)} \Phi\left(u; \psi(S_T)\right) du,$$

where $\phi(u; L_t)$ is the characteristic exponent for the Lévy process. Therefore, the Fourier transform for the value of the option, $\Phi(u; P(S,t))$, is equal to

$$\Phi(u; P(S,t)) = e^{[-r-iu(r-\omega_S)+\phi(u;L_t)](T-t)} \Phi\left(u; \psi(S_T)\right). \qquad (6.18)$$

One can show that equation (6.18) is a solution for the following partial differential equation

$$\frac{\partial \Phi(u; P(S,t))}{\partial t} = [r + iu(r - \omega_S) - \phi(u; L_t)] \Phi\left(u; \psi(S_T)\right), \qquad (6.19)$$

subject to $\Phi(u; P(S,T)) = \Phi(u; \psi(S_T))$.

By replacing the characteristic exponent for the log-stable, CGMY, and KoBoL processes in equation (6.19), and taking the inverse Fourier transform, the fractional partial differential equation for those processes can be obtained. More specifically, substituting equation (6.17) into equation (6.19) and taking the inverse Fourier transform of both sides gives the following pricing fractional partial differential equation for the log-stable process

$$\frac{\partial P(S,t)}{\partial t} + (r - \omega_S)\frac{\partial P(S,t)}{\partial S} - \omega_S D_{-\infty}^{\alpha} P(S,t) = rP(S,t),$$

where $D_{-\infty}^{\alpha}$ is the left-side RiemannLiouville (RL) fractional derivative with respect to S,[33] and

$$\omega_S = \phi(i; LS),$$
$$= \frac{1}{2}\sigma^{\alpha} \sec\left(\frac{\alpha\pi}{2}\right).$$

Similarly, the fractional partial differential equation for the CGMY is given by

$$\frac{\partial P(S,t)}{\partial t} + (r - \omega_S)\frac{\partial P(S,t)}{\partial S} + C\Gamma(-Y)e^{MS} D_{\infty}^{Y}\left(E^{-MS}P(S,t)\right)$$
$$C\Gamma(-Y)e^{-GS} D_{-\infty}^{Y}\left(E^{GS}P(S,t)\right) = (r - C\Gamma(-Y)(M^Y + G^Y))P(S,t),$$

where D_{∞}^{Y} and $D_{-\infty}^{Y}$ are the right-side and left-side RL fractional derivative,

---

[33] The RL fractional derivative and other definitions for the fractional derivative can be found in Chapter 1.

respectively, and,

$$\omega_S = \phi(-i; CGMY),$$
$$= C\Gamma(Y)\Big((M-1)^Y - M^Y + (G+1)^Y - G^Y\Big).$$

Finally, the fractional partial differential equation for a European option, when the dynamics of the asset price process is the KoBoL process is

$$\frac{\partial P(S,t)}{\partial t} + (r - \omega_S - \lambda^{\alpha-1}(q-p))\frac{\partial P(S,t)}{\partial S}$$
$$+ \frac{1}{2}\sigma^\alpha\Big(pe^{\lambda S}D_\infty^\alpha e^{-\lambda S}P(S,t) + e^{-\lambda S}D_{-\infty}^\alpha e^{\lambda S}P(S,t)\Big) = \Big(r - \frac{1}{2}\sigma^\alpha\lambda^\alpha\Big)P(S,t),$$

where

$$\omega_S = \phi(-i; KoBoL),$$
$$= \frac{1}{2}\sigma^\alpha\Big(p(\lambda-1)^\alpha + q(\lambda+1)^\alpha - \lambda^\alpha - \alpha\lambda^{\alpha-1}(q-p)\Big).$$

The analytic solution for the fractional partial differential equation of the log-stable, CGMY, and KoBoL processes is not available. However, by using numerical approaches such as the finite differences, finite elements, spectral methods (Galerkin and collocation), homotopy perturbation method, variational iteration method, and Adomian decomposition method one can obtain the numerical approximation for those fractional partial differential equations.[34]

By applying the same approach for the stable process and all classes of the tempered stable process, one can obtain the related fractional partial differential equations. As we discussed in Chapter 2, another application of a fractional partial differential equation in financial economics is providing tools for obtaining the numerical approximation for the probability density function for the stable, log-stable, and all classes of tempered stable processes. In Chapter 2, we applied this approach in the case where t is equal to 1. However, one can extend this approach for any value for t.

The fractional partial differential equation for the simplest option, a European call and put option, discussed above can be extended to other types of options. For example, since an American option is exercisable at any point in time prior to maturity, instead of an equality in the fractional partial differential equation, there is an inequality (i.e., $\leq$ or $\geq$).

Key points of the chapter

• The Black-Scholes formula for option pricing has several drawbacks.

---

[34] More information about the application of the numerical approach to the fractional partial differential equation can be found in Meerschaert and Tadjeran (2004), Meerschaert and Tadjeran (2006), Carr and Mayo (2007), Fallahgoul et al. (2012b), and Fallahgoul (2013). The homotopy perturbation, variational iteration, and Adomian decomposition methods are discussed in Chapter 2.

- The Black-Scholes formula does not take into account two stylized facts about historical returns: heavy tails and excess kurtosis, as well as assuming a constant volatility.
- There is a one-to-one relation between volatility and the option price.
- Given the market price of an option, one can obtain volatility which is referred to as implied volatility.
- The constant volatility in the Black-Scholes formula is a consequence of the flat implied volatility curve.
- In the Black-Scholes formula it is assumed that the dynamics of the asset price process follows a geometric Brownian motion.
- The Lévy process provides a rich class of stochastic processes with useful properties for application in financial economics.
- By replacing the geometric Brownian motion in the Black-Scholes model with the Lévy process such as CGMY, variance-gamma, or normal inverse Gaussian, a more realistic value for an option can be obtained.
- A Lévy process can include a finite or infinite number of jumps.
- Different stochastic processes are introduced as an alternative to geometric Brownian motion.
- The CGMY and KoBoL processes belong to the class of tempered stable processes.
- Brownian motion, stable, log-stable, CGMY, KoBoL, and all classes of tempered stable processes are subclasses of the Lévy process.
- Stable, log-stable, CGMY, and KoBoL processes capture the empirical properties observed for financial time series.
- There are three different approaches for obtaining the value of an option: stochastic differential equation, partial differential equation, and fractional partial differential equation.
- Because of the existence of jumps, fractional partial differential equations are obtained rather than partial differential equations for option pricing.
- There are different approaches for solving numerically a partial differential equation and fractional partial differential equation: finite differences, finite elements, spectral method.
- The finite difference method is the most popular model for solving a partial differential equation and fractional partial differential equation.

Continuous-Time Random Walk and Fractional Calculus

Continuous-time random walk is an extension of the random walk. More specifically, it is constructed by introducing a new source of randomness to the random walk. This new source of randomness is waiting time. In this chapter, we first discuss the continuous-time random walk, and then move on to its applications in financial economics.

In Section 7.1, the definition and properties of the continuous-time random walk processes are discussed. Section 7.2 is devoted to the connection of continuous-time random walk processes with fractional derivatives. The continuous-time random walk processes are used for modeling the behavior of dynamics of high frequency asset prices as well as the rate of growth of a firm in Section 7.3.

7.1 Continuous-Time Random Walk

Introduced by Montroll and Weiss (1965), the principal difference between continuous-time random walk and random walk is that the time between two jumps in each step of a random walk is a random variable. This random variable is called the waiting time random variable.

Let $\{t_0 < t_1 < \cdots < t_n\}$ be $n+1$ points of time. Consider an individual, X_t, who starts to walk at time t_0. The individual, who we will refer to as the "walker", changes position at time t_1 and jumps by a length equal to ΔX_{t_1}. After waiting time j_{t_2}, the walker changes position and jumps by an amount equal to ΔX_{t_1}, and so on. Let $j_i = t_{i+1} - t_i$ be the waiting time random variable for changing the walker's position. More specifically,

$$j_1 = t_1 - t_0,$$
$$j_2 = t_2 - t_1,$$
$$\vdots$$
$$j_i = t_{i+1} - t_i,$$
$$\vdots$$

Then the walker's position at time $t - n$ is given by

$$X_{t_n} = \Delta X_{t_0} + \Delta X_{t_1} + \cdots + \Delta X_{t_n} = \sum_{i=0}^{N(t)} \Delta X_{t_i},$$

http://dx.doi.org/10.1016/B978-0-12-804248-9.50007-3,

or equivalently

$$X_{t_n} = X_{T_n},$$

where $\Delta X_0 = X_0 = 0$, $T_n = j_0 + j_1 + \cdots + j_n$. The random variable X_{t_n} is called the total displacement of the walker at time t_n and it is referred to as the jump random variable, and $N(t)$ is the random number of jumps defined as follows

$$N(t) = max\{n; T_n \leq t\}.$$

The waiting time random variables, $(j_i)_{i=0}^{\infty}$, are independent and identically distributed variables (mutually independent). There are two possibilities for the relationship between random variables T_n and X_t. They can be either independent or correlated.

The probability density function for the walker being at position X at time t provides a useful tool for studying the continuous-time random variable. Let $f(\Delta X, j)$ be joint probability density function of the jump variable and waiting time. Then the marginal probability density function is given by

$$g(j) = \int_{-\infty}^{\infty} f(\Delta X, j) d\Delta X,$$

$$h(\Delta X) = \int_{0}^{\infty} f(\Delta X, j) dj,$$

and

$$G(j) = 1 - \int_{0}^{t} g(s) ds.$$

Let $f(X_{t_i}, s)$ be the probability density function for the walker being at position $X_{t_{i+1}}$ at time t_{i+1}, then

$$f(X_{t_{i+1}}, t_{i+1}) = \delta(X_{t_{i+1}}) G(t_{i+1}) + \int_{t_i}^{t_{i+1}} \int_{-\infty}^{\infty} f(X_{t_{i+1}} - X_{t_i}, t_{t_{i+1}} - t_i) f(X_{t_i}, t_i) dX_{t_i} dt_{t_i},$$

$$(7.1)$$

where $\delta(X_{t_{i+1}})$ is the Dirac's delta function and $f(X_{t_i}, s)$ is known.[1] Bear in mind that the Poisson and compound Poisson processes are a continuous-time random variable where the waiting times are a constant and an exponential random variable, respectively. Therefore, these two processes belong to the class of Lévy processes. In fact, they have stationary and independent increments, and their distributions are an infinite divisible distribution.[2]

Equation (7.1) is an integral equation. By solving it, one obtains the probability density function $f(X, t)$. The best way for solving integral equation (7.1) is by using the Laplace and Fourier transform and using limit theorems. Laplace and

---

[1] $f(X_{t_i}, t_i)$ is the probability density function for the walker being at position X_{t_i} at time t_i.
[2] Detailed information about the infinite divisible distribution and the Lévy process can be found in Rachev et al. (2011) and Cont and Tankov (2003).

Fourier transforms are given by

$$f(u) = L\{f(x)\} = \int_0^\infty e^{-ux} f(x) dx,$$

$$F(\omega) = F\{f(x)\} = \int_{-\infty}^\infty e^{i\omega x} f(x) dx.$$

It can be shown that[3]

$$L\{F\{f(X,t)\}\} = \frac{L\{G(t)\}}{1 - L\{F\{f(X,j)\}\}}. \tag{7.2}$$

In order to obtain probability density function $f(X,t)$ from equation (7.2), one has to calculate the inverse of Laplace and Fourier transforms. The operator $L\{F\{.\}\}$ is called the Laplace-Fourier transform. If the jumps and waiting time are independent, then the solution for integral equation (7.1) exists.[4]

7.2 Fractional Calculus and Probability Density Function

In the previous section, we show that the probability density function for the continuous-time random variable is the solution to the integral equation (7.1). In this section, we use fractional derivatives and integrals for obtaining the solution to the integral equation (7.1). More specifically, by using the Caputo fractional derivative,[5] we construct a new equivalent problem and then the solution is provided.

Obtaining the solution to the integral equation (7.1) can be divided into two cases: coupled and uncoupled case. For the uncoupled case

$$f(\Delta X, j) = h(\Delta X)g(j), \tag{7.3}$$

and in the coupled case $f(\Delta X, j)$

$$f(\Delta X, j) = H(h(\Delta X), g(j))$$

where function H cannot be separated such as equation (7.3).[6]

---

[3] Detailed information may be found in Davies and Martin (1979).

[4] More information about the procedure for solving integral equation (7.1) can be found in Scalas (2006) and references therein.

[5] See Chapter 1 for an explanation of the Caputo fractional derivative.

[6] Detailed information about the uncoupled case can be found in Scalas et al. (2004).

7.2.1 Uncoupled

Let $f(\Delta X, j) = h(\Delta X)g(j)$. Since g and h are probability density functions, they must satisfy the following[7]

$$\int_0^\infty g(s)ds = 1,$$

$$\int_{-\infty}^\infty h(x)dx = 1.$$

For simplicity, we set $t_i = 0$ and $t_{i+1} = t$. It can be shown that the integral equation (7.1) is equivalent to

$$f(X, t) = \delta(X)G(t) + \int_0^t \int_{-\infty}^\infty f(X - u, t - s)f(u, s)duds,$$

$$= \delta(X)G(t) + \int_0^t g(t - s)\left(\int_{-\infty}^\infty h(X - u)f(u, s)du\right)ds. \qquad (7.4)$$

Let

$$h_n(X) = \int_{-\infty}^\infty \cdots \int_{-\infty}^\infty h(X - u_{n-1})h(X - u_{n-2}) \cdots h(u_1)du_{n-1}du_{n-2} \cdots du_1,$$

be an n−fold convolution of the jump density. The probability of n jumps up to time t is given by

$$F(n, t) = \int_0^t g_n(t - j)G(j)dj$$

where $g_n(j)$ is an n−fold convolution of the waiting time density

$$g_n(j) = \int_{-\infty}^\infty \cdots \int_{-\infty}^\infty h(j - j_{n-1})h(j - j_{n-2}) \cdots h(j_1)dj_{n-1}dj_{n-2} \cdots dj_1.$$

Therefore, by using the Laplace and Fourier transforms, the solution to the integral equation (7.4) is given by

$$f(X, t) = \sum_{n=0}^\infty F(n, t)h_n(X).$$

Now, by using the Caputo fractional derivative, $F(n, t)$ can be calculated. More precisely, the Caputo fractional derivative of $G(j)$ is given by

$$\frac{\partial^\alpha G(j)}{\partial j^\alpha} = -G(j) \qquad (7.5)$$

where $0 < \alpha \le 1$, $j > 0$, $G(0^+) = 1$, and $\dfrac{\partial^\alpha}{\partial j^\alpha}$ is the Caputo fractional derivative.

---

[7] See, Scalas et al. (2004).

By applying the Laplace transform to equation (7.5), the Laplace transform of $G(j)$ is given by

$$L\{G(j)\} = \frac{u^{\alpha-1}}{1 + u^\alpha}. \tag{7.6}$$

By inverting the Laplace transform (7.6), the following solution to equation (7.5) is obtained

$$G(j) = E_\alpha(-j^\alpha),$$

where

$$E_\alpha = \sum_{n=0}^{\infty} \frac{z^n}{\Gamma(\alpha n + 1)}.$$

In physics, E_α is called the Mittag-Leffler function. It is a possible model for a heavy tail survival function.

One can show that the Mittag-Leffler function for small j is equal to

$$G(j) = E_\alpha(-j^\alpha) \simeq 1 - \frac{j^\alpha}{\Gamma(\alpha + 1)} \simeq e^{\frac{-j^\alpha}{\Gamma(\alpha + 1)}},$$

$$g(j) \simeq \frac{j^{-(1-\alpha)}}{\Gamma(\alpha)}.$$

However, for large values of j, $j \to \infty$, the Mittag-Leffler function has power-law asymptotic representation. That is,

$$G(j) = E_\alpha(-j^\alpha) \simeq \frac{\Gamma(\alpha)\sin(\alpha\pi)}{\pi j^\alpha},$$

$$g(j) \simeq \frac{\Gamma(\alpha + 1)\sin(\alpha\pi)}{\pi j^{\alpha+1}}.$$

Basically one can show that there is a one-to-one relation between continuous-time random walks and fractional diffusion. More information may be found in Scalas et al. (2004) and Scalas (2006).

Therefore, the solution for the integral equation (7.1) is given by

$$f(X, t) = \sum_{n=0}^{\infty} E_\alpha^{(n)}\left(-t^\alpha\right)\frac{t^{\alpha n} h_n(X)}{n!},$$

where

$$E_\alpha^{(n)}(z) = \frac{\partial^n E_\alpha}{\partial z^n}.$$

7.2.2 Coupled Case

The coupled case for the continuous-time random walk was studied by Meerschaert and Scheffler (2001) and Meerschaert et al. (2002). The results obtained in these two papers are summarized in the following theorem and corollary.

Theorem 7.1 *(Scalas (2006) and Meerschaert and Scheffler (2001)) Let $f(\Delta X, j)$ be joint density function for a continuous-time random walk. If under under scaling $\Delta X \to a\Delta X$ and $j \to bj$, the Fourier and Laplace transform of $f(\Delta X, j)$ behaves as follow*

$$L\{F\{f_{a,b}(\Delta X, j)\}\} = L\{F\{f(a\Delta X, bj)\}\},$$

and if $a \to 0$ and $b \to 0$, the asymptotic relation holds

$$L\{F\{f_{a,b}(\Delta X, j)\}\} = L\{F\{f(a\Delta X, bj)\}\} \simeq 1 - \mu|a\omega|^{\alpha} - \nu(bu)^{\beta},$$

with $0 \leq \alpha \leq 2$ and $0 < \beta \leq 1$, then under scaling relation $\mu a^{\alpha} = \nu b^{\beta}$, the solution of the scaled coupled continuous-time random walk integral equation (7.1) weakly converges to the Green function of the fractional diffusion equation, $u(X, t)$, for $a \to 0$ and $b \to 0$.

Proof See Scalas (2006). □

If the first moment of the waiting time random variable and the second moment of the jump random variable are finite, then for the coupled case the probability density function for the integral equation (7.1) is the solution of an ordinary diffusion equation. The following corollary shows this consequence of Theorem 7.1.

Corollary 7.1 *If the Laplace-Fourier transform of $f(\Delta X, j)$ is regular for $\omega = 0$ and $u = 0$, and, the marginal waiting-time density has finite first moment and the marginal jump density is symmetric with finite second moment, then the limiting solution of integral equation (7.1) for the coupled continuous-time random walk is the Green function of the ordinary diffusion equation.*

Proof See Scalas (2006). □

7.3 Applications

In this section, applications of continuous-time random walk in financial economics are discussed. More specifically, we use the continuous-time random walk process for modeling dynamics of asset prices.

7.3.1 Dynamics of the Asset Prices

Modeling dynamics of asset prices plays important role in a lot of microeconomics problems. For example, by understanding the behavior of stock prices, one can take good decision for a portfolio. Continuous-time random walk process is a suitable class of process for modeling the behavior of high frequency data.

Figure 7.1 shows a trajectory for the continuous-time random walk. It shows two random variables play important role in the structure of this process: jump magnitude, and waiting time. Unlike the random walk, the waiting time for jumps is not the same during time. It is perfect to describe the behavior of dynamics of high frequency market. More precisely, since the business activity for all trading periods are not the same.[8] Some days we have more activity and some days less. The waiting time random variable describes this property.

In the continuous-time random walk setting the calender time, t, is replaced with business time, waiting time random variable. Clark (1973) for the first time tried to recover normality assumption of distribution for the time series of return process. His ideas are applied to the option pricing with pure jump Lévy process. These pure jump Lévy process can be obtained by changing the time of a Brownian motion with a subordinator. Variance-gamma, normal inverse Gaussian, and CGMY process are pure jump Lévy process that can be obtained by changing time of Brownian motion with a subordinator.[9]

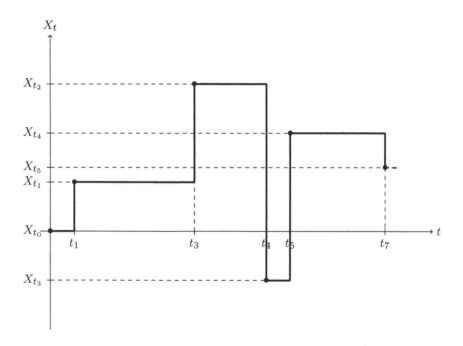

Figure 7.1 A trajectory for the continuous-time random walk process: tick-by-tick price fluctuation.

The initial setting for implementing the continuous-time random walk for studying the behavior of asset prices is as follows. Denote the waiting times between each trade by $\{j_1, j_2, \cdots, j_n, \cdots\}$. Let $(P_t)_{t \geq 0}$ and $S_t = \log(P_t)$ be the price process and log-price of an asset at time t. The waiting time random variables are independent and identically distributed. Let $X_1, X_2, \cdots, X_n, \cdots$ be the

---

[8] These periods are 5 minutes, 10 minutes, 30 minutes, and 1 hour. Moreover, it can be extended for modeling the dynamics of the low frequency asset prices such as daily, weekly, and monthly.

[9] Detailed information about this approach and Variance-gamma, normal inverse Gaussian, and CGMY process can be found in Cont and Tankov (2003).

log-return process, more specifically, X_n is given by

$$X_n = S_n - S_{n-1} = \log(P_n) - \log(P_{n-1}) = \log\left(\frac{P_n}{P_{n-1}}\right).$$

Without loss of generality we assume $X_0 = 0$. Bear in mind that the log-return is more convenient to study the behavior of asset price.[10] Moreover, the log-return random variables are independent and identically distributed.

As explained earlier, the continuous-time random walk process is classified into two cases: uncoupled and coupled. If the random variables j_n and ΔX_n are independent for each value of n, then X_n is called an uncoupled continuous-time random walk process, and if they are dependent, X_n is called a coupled continuous-time random walk process. Let $T_n = j_1 + j_2 + \cdots + j_n$ be the time nth trade. The number of trades by time $t > 0$ is $N_t = max\{n : T_n \leq t\}$, and therefore, the log-price at time t is given by

$$S_{N_t} = \log(P_{T_n}) = X_1 + X_2 + \cdots + X_{N_t}. \tag{7.7}$$

Equation (7.7) is a subordinated process. More specifically, the calender time, t, for the stochastic process S_t is changed with business time, N_t. If the waiting times are exponentially distributed, the continuous-time random walk process is a compound Poisson process. Therefore, the continuous-time random process is a Markovian process that belongs to the class of Lévy processes. In this case, the distribution for the log-price is Gaussian and for the price the distribution is log-normal.

The asymptotic theory for the continuous-time random walk process provides a useful tool for the application of this process. In fact, in order to estimate the parameter of interest one should estimate parameters of asymptotic distributions. In the previous section we discussed an approach for obtaining the probability density function for a continuous-time random walk for both the coupled and uncoupled cases.

Some properties for the distribution for a continuous-time random walk process are summarized as follows:

- If the log-returns process has finite variance and $c \to \infty$, then $c^{-\frac{1}{2}} S_{ct} \to B_t$. B_t is Brownian motion and its probability density function, $f(X, t)$, is the solution of

$$\frac{\partial f(X, t)}{\partial t} = D \frac{\partial^2 f(X, t)}{\partial X^2}, \tag{7.8}$$

where $D > 0$.

- If the waiting time random variables have a finite mean such as $\frac{1}{\lambda}$, then

$$N_t \sim \lambda t, \quad \text{when} \quad t \to \infty.$$

Therefore, the the scaling limit of a continuous-time random walk is a Brownian motion and its probability density function satisfies equation (7.8).

---

[10] Benefits of using log-returns are log-normality and time-additivity. More information about these concepts may be found in Tsay (2005).

- If the distribution of log-returns is symmetric, with zero mean, and its tail has power-law probability, then the random walk S_n is asymptotically a stable process. More specifically, if

$$P(|X_n| > r) \sim r^{-\alpha},$$

then

$$c^{-\frac{1}{\alpha}} S_{ct} \to A_t,$$

where A_t is a stable process. The probability density function for the stable process A_t is the solution

$$\frac{\partial f(X, t)}{\partial t} = D \frac{\partial^\alpha f(X, t)}{\partial |X|^\alpha}.$$

- If the continuous-time random walk is uncoupled, the waiting times and log-returns random variables are independent,[11] then the probability density function for the continuous-time random walk process is the solution of

$$\frac{\partial^\beta f(X, t)}{\partial t^\beta} = D \frac{\partial^\alpha f(X, t)}{\partial |X|^\alpha} + \frac{\delta(X) t^{-\beta}}{\Gamma(1 - \beta)}.$$

The continuous-time random walk process is used for the stock and future markets in Mainardi et al. (2000) and Raberto et al. (2002), where the unconditional distribution for the waiting time random variable is non exponential. Scalas (2006) provides a survey on application of the continuous-time random walk process for the foreign-exchange market, the futures market for German government bonds, and German government bond futures contracts.

Key points of the chapter

- The continuous-time random walk is an extension of the discrete random walk process.
- The difference between the continuous-time random walk and the discrete random walk processes is related to the waiting time to the next jump.
- In the continuous-time random walk process the waiting time random variables are independent and identically distributed.
- The probability density function for the continuous-time random walk process satisfies an integral equation.
- The probability density function for the continuous-time random walk process does not exist in closed-form, it can be obtained in an asymptotic form.
- Based on the correlation between waiting time random variables and jump variables, the continuous-time random walk process is divided into two cases: coupled, and uncoupled.
- Continuous-time random walk processes are used to model the dynamics of asset prices.

[11] Detailed information about properties for the coupled case can be found in Meerschaert and Scalas (2006).

• The compound poisson process is special class of the continuous-time random walk processes where the distribution of the waiting time random variable is exponential.

Applications of Fractional Processes

Chapter 2 described how fractional calculus can be applied to generate fat-tailed distributions; Chapter 6 discussed how to apply fractional processes to the pricing of derivatives. As fractional processes are not semi martingales, violations of the no-arbitrage condition might occur. We have seen how to circumvent this problem.

This chapter discusses financial applications other than option pricing where fractional processes are used. In fact, fractional processes can be applied to many processes that have long memory. We will therefore begin by discussing long-memory processes of interest in finance. These might be stationary processes with long memory as opposed to short memory, but it might also include non stationary processes with long memory, though not the infinite memory of random walks.

The usual unit root tests tend to conclude that many time series in finance and economics have a unit root. However in many instances, such as interest rates, there are economic reasons to believe that unit root behavior might not be a plausible model. In these cases, fractionally integrated processes might provide a viable alternative. Testing for long memory is intrinsically difficult because it involves testing autocorrelations over long time horizons. To overcome this problem, a number of researchers are exploring self-similarity; their objective is to use the behavior at short time horizons to estimate behavior at long time horizons. However, self-similarity and long-range memory are not equivalent properties. Non Gaussian Levy processes might exhibit self-similarity but no long memory, while there are examples of processes with long memory but no self-similarity. Cont (2005) provides a concise summary of the key findings and theoretical considerations.

Given the above remarks, what are the processes that might have long memory? In the literature, the following processes have been considered:

1. stock returns and volatility
2. interest rates and currency rates
3. order arrivals

In order to discuss the modeling of these processes, we have first to introduce fractionally integrated discrete-time processes, that is, fractionally integrated time series.

http://dx.doi.org/10.1016/B978-0-12-804248-9.50008-5,

8.1 Fractionally Integrated Time Series

Fractionally integrated time series was introduced operationally in Granger (1980), Granger and Joyeux (1980), and Hosking (1981). Let's recall that an autoregressive moving-average model of integrated time series (ARIMA) is written as follows:

$$a\left(L\right)\left(1-L\right)y_t = b(L)\varepsilon_t$$

where $a\left(L\right), b(L)$ are polynomials in the lag operator L and ε_t is a zero-mean serially uncorrelated process. A fractionally integrated process is written as:

$$a\left(L\right)\left(1-L\right)^d y_t = b(L)\varepsilon_t, -0.5 < d < 0.5$$

The fractional differencing operator $\left(1-L\right)^d$ admits an infinite expansion:

$$\left(1-L\right)^d = \sum_{k=0}^{\infty} \Gamma\left(k-d\right)\Gamma\left(k+1\right)^{-1}\Gamma\left(-d\right)^{-1} L^k$$

For $-0.5 < d < 0.5$ the process y_t is stationary and invertible and for $0 < d < 0.5$ the process has long memory in the sense that the sum $\sum_{i=-k}^{k} |\rho_i|$, where ρ_i is the autocorrelation at lag i does not converge when k tends to infinity.

The definition of fractional time series given above is based on an infinite moving average of noise terms. Johansen (2008) described a Vector Autoregressive Model whose solution is a fractionally integrated time series and which also allows for fractional cointegration. Granger (1986) had already suggested a notion of fractional cointegration, proposing the following model:

$$a^*(L)(1-L)^d y_t = (1-(1-L)^d)(1-L)^{d-b}\alpha\beta' y_t + b(L)\epsilon_t$$

where ϵ_t is a sequence of i.i.d variables with positive definite covariance matrix.

Johansen (2008) proposed a different VAR model. Lets define the "lag operator" $L_d = (1-(1-L)^d)$ which, as observed already in Granger (1986), plays a role similar to the lag operator. The model proposed in Johansen (2008) is written as:

$$A(L_d)(1-L)^d y_t = (1-(1-L)^d)(1-L)^{d-b}\alpha\beta' y_t + b(L)\epsilon_t$$

In the Johansen (2008) model the polynomial A is defined on the lag operator L_d while in the Granger (1986) model the polynomial a^* was defined on the usual lag operator L.

The Johansen model is not a fractional ARIMA model which is defined as:

$$D(L)(1-L)^d y_t = B(L)\epsilon_t$$

where $D(L)$ and $B(L)$ are finite order lag polynomials. It is, however, a fractional ARIMA model in the lag operator L_d.

Fractional cointegration means that, given a multivariate process y_t with fractional integration d there are linear combinations of y_t which have a lower level of integration $d - b, b > 0$.

8.2 Stock-Returns and Volatility Processes

The question of the long memory of stock returns has a long history. Mandelbrot (1971) first suggested that returns exhibit a long-range memory. Initial tests based on rescaled range (R/S) analysis seemed to prove that there is effectively some long-range memory of returns. Lo (1991) criticized tests based on R/S analysis, observing that these tests were very sensitive to the short-term component; he introduced a modified R/S test. Using this modified test, Lo reached a negative conclusion. But it was later observed in Teverovsky et al. (1999) that Lo's modified R/S test was too conservative. The authors find evidence of long range memory in the CRSP files but they find that the low value of the Hurst exponent (H = 0,6) makes this conclusion uncertain.

Willinger et al. (1999) caution against using one single test to reach a conclusion regarding long-range memory. Ultimately, using a battery of tests, these authors conclude that there is not much evidence that returns have long-memory properties. They found that the decay of autocorrelation of individual stock returns was very rapid. However, absence of autocorrelation of returns does not imply absence of autocorrelation of non linear functions of returns. In particular, the absolute value of returns and power functions of absolute returns show strong autocorrelation. In other words, the sign of returns is basically unpredictable but the magnitude of returns is predictable.

This phenomenon is related to the well-known *clustering of volatility*, which means that extended periods of high volatility are followed by extended periods of low volatility. Clustering, however, is not equivalent to autocorrelation. Empirically it was found that the autocorrelation function of volatility decays as a power law.

The first attempt to model volatility was made by Engle (1982), with the introduction of the autoregressive model of conditional volatility now known as the autoregressive conditional heteroscedasticity (ARCH) model. Engle demonstrated that the ARCH model can represent a stationary process where, however, conditional volatility is time-dependent.

Suppose returns r_t are represented by a factor model

$$r_t = \mu + \beta_1 f_{1t} + \cdots + \beta_k f_{kt} + u_t$$

where the $\beta_i, i = 1, \ldots, k$ are constant factor loadings, $f_{it}, i = 1, \ldots, k$ are the factors and u_t are the residuals which are normally distributed with conditional variance σ_t^2

The ARCH(p) model is written as:

$$\sigma_t^2 = \alpha_0 + \alpha_1 u_{t-1}^2 + \cdots + \alpha_p u_{t-p}^2$$

It is a deterministic model of volatility in the sense that volatility is a deterministic function of past errors.

ARCH models became very popular and a large family of ARCH and GARCH (the G stands for *Generalized*) models were created. The GARCH(p,q) model is written as:

$$\sigma_t^2 = \alpha_0 + \alpha_1 u_{t-1}^2 + \cdots + \alpha_p u_{t-p}^2 + \beta_1 \sigma_{t-1}^2 + \cdots + \beta_q \sigma_{t-q}^2$$

In all these models, parameters are subject to restrictions so that the model represents a stationary process. For example, the GARCH(1,1) model requires that $\alpha_1 + \beta_1 < 1$.

A fractional version of GARCH, called FIGARCH, was proposed by Baillie et al. (1996). To demonstrate how to write a FIGARCH model, let's rewrite the GARCH(p,q) model as follows:

$$\phi(L)(1-L)u_t^2 = \alpha_0 + [1 - \beta(L)]\nu_t$$
$$\nu_t = u_t^2 - \sigma_t^2$$

The FIGARCH model is obtained by replacing $(1-L)$ with $(1-L)^d, 0 < d < 1$. For $0.5 \leq d < 1$ the differencing operator produces a nonstationary process; for $d = 1$ we obtain the integrated GARCH (or IGARCH) process.

A different approach to modeling the stochastic behavior of volatility is stochastic volatility. Stochastic Volatility (SV) models consider volatility as a separate process. Different specifications of the volatility process are possible. Widely used specifications are the Hull and White (1987) model and the model developed in Heston (1993). The Heston model adopts a mean-reverting model for volatility written as follows:

$$dS_t = S_t(\mu dt + \sigma_t dB_t)$$
$$d(\sigma_t^2) = -\kappa(\sigma_t^2 - \sigma_0^2)dt + \gamma \sigma_t dB'_t, \langle dB dB' \rangle = \rho < 0$$

However, as noted earlier, empirical studies have shown that the decay of autocorrelation of volatility follows a power law. In order to obtain a decay of volatility closer to the power laws that are found empirically, it was suggested to model volatility with fractional processes. The first continuous-time fractional-volatility model was that of Comte and Renault (1998):

$$dS_t = S_t(\mu dt + \sigma_t dB_t)$$
$$d(\log \sigma_t) = -\kappa(\vartheta - \log \sigma_t)dt + \gamma \sigma_t dB'_t, \langle dB dB' \rangle = \rho < 0$$

followed by Comte et al. (2012). A more recent specification is the Mendes et al. (2015) fractional volatility model which is written as follows:

$$dS_t = (\mu_t S_t dt + \sigma_t S_t dB_t)$$
$$d(\log \sigma_t) = \beta + \frac{k}{\delta}\{B_H(t) - B_H(t-\delta)\}k\kappa(\vartheta - \log \sigma_t)dt$$
$$+ \gamma \sigma_t dB'_t, \langle dB dB' \rangle = \rho < 0$$

In summary, empirical studies suggest that volatility exhibits a slow power-law decay of the autocorrelation function. To capture this phenomenon, a number of stochastic volatility models with fractional processes representing volatility has been proposed.

8.3 Interest-Rate Processes

A number of empirical studies tend to conclude that interest rates have long memory. Goliński and Zaffaroni (2016) neatly summarize the issue as follows: "One of the main challenges in modelling the term structure of interest rates is the fact that nominal observed yields are extremely persistent. In fact, they are essentially non distinguishable from a non stationary series: any test would hardly reject the hypothesis of a unit root."

However, the assumption of a unit root has implausible economic consequences. For this reason, fractional processes have been proposed as a suitable model for interest rates.

The use of fractional processes to represent interest rates was first suggested by Backus and Zin (1993) and by Duan and Jacobs (1996). Comte and Renault (1996) proposed a model of the term structure of interest rates in continuous time. More recently several studies have proposed fractionally integrated processes for modeling interest rates in the framework of affine models.

Osterrieder et al. (2013) proposed fractional cointegration between factors in an affine model of interest rates. Goliński and Zaffaroni (2016) proposed a general theory of fractional processes to model affine term structures of interest rates. The authors consider an affine model where short-term rates are represented by factor models. Factors obey a fractionally integrated ARFIMA model. This model represents the long-term persistence empirically found in interest rates. In addition, it admits an infinite-dimensional state-space representation which can be conveniently estimated with Kalman filters.

8.4 Order Arrival Processes

An important area where fractional processes find application is the modeling of the order arrival process. This topic has become ever more interesting with the diffusion of high- and ultrahigh-frequency data and with high-frequency trading. The availability of data has enabled the empirical study of the econometrics of high frequency data while high frequency trading has provided a powerful economic motivation for their study.

Ultrahigh-frequency data are tick-by-tick data. As orders and trading occur at random times, tick-by-tick data are not classical time series but point processes. The representation of tick-by-tick data includes a representation of the order-arrival process and a representation of the magnitude of the transaction.

Scalas et al. (2000) and Mainardi et al. (2000) were the first to apply tick-by-tick data to the formalism of Continuous-Time Random Walks (CTRW) developed in Montroll and Weiss. A CTRW is a random walk where both the length of the time between two steps and the magnitude of each step are random. Call t_i, x_i the time and magnitude of the i-th transaction, call $\tau_i = t_{i+1} - t_i, \xi_i = x_{i+1} - x_i$ the waiting times and the magnitude of jumps, and call $\varphi(\xi, \tau)$ the joint density of jumps and waiting times and $p(x, t)$ the joint probability that the diffusing quantity be at position x at time t. Montroll and Weiss (1965) demonstrated that the Laplace-Fourier transform of p has the following form:

$$\tilde{p}(\kappa, s) = \frac{1 - \tilde{\psi}(s)}{s} \frac{1}{1 - \tilde{\varphi}(\kappa, s)}$$

where $\tilde{p}(\kappa, s)$ is the Laplace-Fourier transform of $p(x, t)$; $\tilde{\psi}(s)$ is the Laplace transform of the waiting time pdf $\psi(\tau) = \int \varphi(\xi, \tau) d\xi$.

Scalas et al. (2000) demonstrated that in the hydrodynamic limit (i.e., long waiting times and long jumps), assuming that:

$$\tilde{\varphi}(\kappa, s) = 1 - s^{\beta} - |\kappa|^{\alpha}, 0 < \beta \leq 1, 0 < \alpha \leq 2$$

the density $p(x, t)$ obeys the following fractional partial differential equation:

$$\frac{\partial^{\beta} p(x, t)}{\partial t^{\beta}} = \frac{\partial^{\alpha} p(x, t)}{\partial |x|^{\alpha}} + \frac{t^{-\beta}}{\Gamma(1 - \beta)} \delta(x).$$

Empirical tests of the above formalism in different markets can be found in Scalas et al. (2004), Scalas (2006), Politi and Scalas (2007), Politi and Scalas (2008), and Sazuka et al. (2009).

The formalism of CTRW assumes that the waiting times are sequences of independent and identically distributed random variables. This assumption might be too restrictive. Engle and Russell (1998) introduced the Autoregressive Conditional Duration (ACD) model. The ACD model has been followed by many similar models where conditional duration or some function of duration is modeled as an autoregressive process. However, it has been observed empirically that the decay of the autocorrelation function of the duration process can be very slow. To model this behavior, Jasiak (1999) introduced the Fractionally Integrated Autoregressive Conditional Duration (FIACD) model. The formalism is similar to the FIGARCH described above.

Key points of the chapter

- Many processes of interest in finance exhibit a long-range memory.
- In addition to stock prices and stock returns, interest rates, volatility and waiting time between orders are processes that exhibit long memory.
- Several academic papers have proposed to model these processes with fractionally integrated processes or using fractional differential equations.

References

Andersen, L. and Lipton, A. (2013). Asymptotics for exponential Lévy processes and their volatility smile: survey and new results. *International Journal of Theoretical and Applied Finance* 16: 1350001.

Backus, D. and Zin, S. E. (1993). Long-memory inflation uncertainty: Evidence from the term structure of interest rates. *Journal of Money, Credit and Banking* 25: 681–700.

Bailey, D. H. and Swarztrauber, P. N. (1994). Computing VAR and AVaR in infinitely divisible distributions. *SIAM Journal on Scientific Computing* 5: 1105–1110.

Baillie, R. T. (1996). Long memory processes and fractional integration in econometrics. *Journal of Econometrics* 73: 5–59.

Black, F. and Scholes, M. (1973). The pricing of options and corporate liabilities. *Journal of Political Economy* 81: 637–654.

Boyarchenko, S. I. and Levendorski, S. Z. (2000). Option pricing for truncated Lévy processes. *International Journal of Theoretical and Applied Finance* 3: 549–552.

Brollerslev, T., Chou, R. and Kroner, K. (1992). ARCH modeling in finance. *Journal of Econometrics* 52: 5–59.

Carr, P., Geman, H., Madan, D. B. and Yor, M. (2002). The fine structure of asset returns: An empirical investigation. *Journal of Business* 75: 305–333.

Carr, P. and Mayo, A. (2007). On the numerical evaluation of option prices in jump diffusion processes. *European Journal of Finance* 13: 353–372.

Carr, P. and Wu, L. (2003). The finite moment log stable process and option pricing. *The journal of finance* 58: 753–778.

Cartea, A. and Castillo-Negrete, D. del (2007). Fractional diffusion models of option prices in markets with jumps. *Physica A: Statistical Mechanics and its Applications* 374: 749–763.

Cheridito, P. (2001a). Mixed fractional Brownian motion. *Bernoulli* 7: 913–934.

Cheridito, P. (2001b). Regularizing fractional Brownian motion with a view towards stock price modelling. Ph.D. thesis, Swiss Federal Institute of Technology, Zurich.

Cheridito, P. (2003). Arbitrage in fractional Brownian motion models. *Finance and Stochastics* 7: 533–553.

Cheridito, P. (2004). Gaussian moving averages, semimartingales and option pricing. *Stochastic processes and their applications* 109: 47–68.

Clark, P. K. (1973). A subordinated stochastic process model with finite variance for speculative prices. *Econometrica* 41: 135–155.

Comte, F., Coutin, L. and Renault, É. (2012). Affine fractional stochastic volatility models. *Annals of Finance* 8: 337–378.

Comte, F. and Renault, E. (1996). Long memory continuous time models. *Journal of Econometrics* 73: 101–149.

Comte, F. and Renault, E. (1998). Long memory in continuous-time stochastic volatility models. *Mathematical Finance* 8: 291–323.

Cont, R. (2005). Long range dependence in financial markets. In *E. Lutton and J. Levy Vehel (eds.), Fractals in Engineering*. Springer, 159–179.

Cont, R. and Tankov, P. (2003). *Financial Modelling with Jump Processes, 2*. CRC press.

Davies, B. and Martin, B. (1979). Numerical inversion of the Laplace transform: A survey and comparison of methods. *Journal of Computational Physics* 33: 1–32.

Decreusefond, L. and Üstüne, A. (1999). Stochastic analysis of the fractional Brownian motion. *Potential Analysis* 10: 177–214.

Duan, J.-C. and Jacobs, K. (1996). A simple long-memory equilibrium interest rate model. *Economics Letters* 53: 317–321.

DuMouchel, W. H. (1975). Stable distributions in statistical inference: 2. information from stably distributed samples. *Journal of the American Statistical Association* 70: 386–393.

Engle, R. F. (1982). Autoregressive conditional heteroscedasticity with estimates of the variance of united kingdom inflation. *Econometrica: Journal of the Econometric Society* : 987–1007.

Engle, R. F. and Russell, J. R. (1998). Autoregressive conditional duration: a new model for irregularly spaced transaction data. *Econometrica* : 1127–1162.

Fallahgoul, H., Hashemiparast, S., Fabozzi, F. J. and Kim, Y. S. (2013). Multivariate stable distributions and generating densities. *Applied Mathematics Letters* 26: 324–329.

Fallahgoul, H., Hashemiparast, S., Fabozzi, F. J. and Klebanov, L. (2014). Analytical-numeric formulas for the probability density function of multivariate stable and geo-stable distributions. *Journal of Statistical Theory and Practice* 8: 260–282.

Fallahgoul, H., Hashemiparast, S., Kim, Y. S., Rachev, S. T. and Fabozzi, F. J. (2012a). Approximation of stable and geometric stable distributions. *Journal of Statistical and Econometric Methods* 1: 97–123.

Fallahgoul, H., Hashemiparast, S. M., Kim, Y. S., Rachev, S. T. and Fabozzi, F. J. (2012b). Approximation of stable and geometric stable distributions. *Journal of Statistical and Econometric Methods* 1: 97–123.

Fallahgoul, H. A. (2013). *Analysis of Stable Distribution Based on Fractional Calculus*. PhD Thesis, K. N. Toosi University of Technology.

Fallahgoul, H. A., Kim, Y. S. and Fabozzi, F. J. (2016). Elliptical tempered stable distribution. *Quantitative Finance* : forthcoming.

Fang, F. and Oosterlee, C. W. (2008). A novel pricing method for european options based on Fourier-cosine series expansions. *SIAM Journal on Scientific Computing* 31: 826–848.

Feller, W. (2008). *An Introduction to Probability Theory and its Applications, 2*. John Wiley & Sons.

Goliński, A. and Zaffaroni, P. (2016). Long memory affine term structure models. *Journal of Econometrics* 191: 33–56.

Gorenflo, R., Mainardi, F., Scalas, E. and Raberto, M. (2001). Fractional calculus and continuous-time finance III: the diffusion limit. In *Mathematical Finance*. Springer, 171–180.

Granger, C. W. (1980). Long memory relationships and the aggregation of dynamic models. *Journal of Econometrics* 14: 227–238.

Granger, C. W. (1986). Developments in the study of cointegrated economic variables. *Oxford Bulletin of Economics and Statistics* 48: 213–228.

Granger, C. W. and Joyeux, R. (1980). An introduction to long-memory time series models and fractional differencing. *Journal of Time Series Analysis* 1: 15–29.

Guégan, D. (2005). How can we define the concept of long memory? an econometric survey. *Econometric Reviews* 24: 113–149.

Halpern, D., Wilson, H. B. and Turcotte, L. H. (2002). *Advanced Mathematics and Mechanics Applications Using MATLAB*. CRC press.

Hashemiparast, S. M. and Fallahgoul, H. (2011a). Approximation of fractional derivatives via gauss integration. *Annali dell'Universit di Ferrara* 57: 67–87.

Hashemiparast, S. M. and Fallahgoul, H. (2011b). Approximation of laplace transform of fractional derivatives via clenshaw–curtis integration. *International Journal of Computer Mathematics* 88: 1224–1238.

Haug, E. G. (2007). *The Complete Guide to Option Pricing Formulas*. McGraw-Hill Companies.

He, J.-H. (1999). Homotopy perturbation technique. *Computer Methods in Applied Mechanics and Engineering* 178: 257–262.

Heston, S. L. (1993). A closed-form solution for options with stochastic volatility with applications to bond and currency options. *Review of Financial Studies* 6: 327–343.

Hosking, J. R. (1981). Fractional differencing. *Biometrika* 68: 165–176.

Hull, J. and White, A. (1987). The pricing of options on assets with stochastic volatilities. *Journal of Finance* 42: 281–300.

Hunt, B. R., Lipsman, R. L. and Rosenberg, J. M. (2014). *A Guide to MATLAB: for Beginners and Experienced Users*. Cambridge University Press.

Hurst, H. E. (1951). Long-term storage capacity of reservoirs. *Transactions of the American Society of Civil Engineers* 116: 770–808.

Itkin, A. (2013). Efficient solution of backward jump-diffusion pides with splitting and matrix exponentials. *arXiv preprint arXiv:1304.3159* .

Itkin, A. and Carr, P. (2012). Using pseudo-parabolic and fractional equations for option pricing in jump diffusion models. *Computational Economics* 40: 63–104.

Jasiak, J. (1999). Persistence in intertrade durations. *Available at SSRN 162008* .

Javaheri, A. (2011). *Inside Volatility Arbitrage: The Secrets of Skewness, 317*. John Wiley & Sons.

Johansen, S. (2008). A representation theory for a class of vector autoregressive models for fractional processes. *Econometric Theory* 24: 651–676.

Kilbas, A. A. A., Srivastava, H. M. and Trujillo, J. J. (2006). *Theory and Applications of Fractional Differential Equations, 204*. Elsevier Science Limited.

Kim, Y. S., Rachev, S. T., Bianchi, M. L. and Fabozzi, F. J. (2010). Computing VaR and AVaR in infinitely divisible distributions. *Probability and Mathematical Statistics* 30: 223–245.

Kim, Y. S., Rachev, S. T., Chung, D. M. and Bianchi, M. L. (2009). Modified tempered stable distribution, GARCH models and option pricing. *Probability and Mathematical Statistics* 29: 91–117.

Klebanov, L. B., Maniya, G. and Melamed, J. A. (1984). A problem of zolotarev and analogs of infinitely divisible and stable distributions in a sheme for summing of a random number of random variables. *Teoriya Veroyatnostei i ee Primeneniya* 29: 757–760.

Kozubowski, T. J. and Rachev, S. T. (1994). The theory of geometric stable distributions and its use in modeling financial data. *European Journal of Operational Research* 74: 310–324.

Lo, A. W. (1991). Long-term memory in stock market prices. *Econometrica* 59: 1279–1313.

Ludvigsson, G. (2015). *Numerical Methods for Option Pricing under the CGMY Process*. MSc thesis, Uppsala Universitet.

Mainardi, F., Raberto, M., Gorenflo, R. and Scalas, E. (2000). Fractional calculus and continuous-time finance ii: the waiting-time distribution. *Physica A: Statistical Mechanics and its Applications* 287: 468–481.

Mandelbrot, B. B. (1971). When can price be arbitraged efficiently? a limit to the validity of the random walk and martingale models. *Review of Economics and Statistics* 53: 225–236.

Mandelbrot, B. B. and Van Ness, J. W. (1968). Fractional Brownian motions, fractional noises and applications. *SIAM Review* 10: 422–437.

Mantegna, R. N. and Stanley, H. E. (1995). Scaling behavior in the dynamics of an economic index. *Nature* 376: 46–49.

Meerschaert, M. M., Benson, D. A. and Bäumer, B. (1999). Multidimensional advection and fractional dispersion. *Physical Review E* 59: 5026.

Meerschaert, M. M., Benson, D. A., Scheffler, H.-P. and Becker-Kern, P. (2002). Governing equations and solutions of anomalous random walk limits. *Physical Review E* 66: 060102.

Meerschaert, M. M. and Scalas, E. (2006). Coupled continuous time random walks in finance. *Physica A: Statistical Mechanics and its Applications* 370: 114–118.

Meerschaert, M. M. and Scheffler, H. P. (2001). *Limit Distributions for Sums of Independent Random Vectors: Heavy Tails in Theory and Practice, 321*. John Wiley & Sons.

Meerschaert, M. M. and Tadjeran, C. (2004). Finite difference approximations for fractional advection–dispersion flow equations. *Journal of Computational and Applied Mathematics* 172: 65–77.

Meerschaert, M. M. and Tadjeran, C. (2006). Finite difference approximations for two-sided space-fractional partial differential equations. *Applied Numerical Mathematics* 56: 80–90.

Mendes, R., Oliveira, M. and Rodrigues, A. (2015). No-arbitrage, leverage and completeness in a fractional volatility model. *Physica A: Statistical Mechanics and its Applications* 419: 470–478.

Menn, C. and Rachev, S. T. (2006). Calibrated FFT-based density approximations for α-stable distributions. *Computational Statistics & Data Analysis* 50: 1891–1904.

Merton, R. C. (1973). Theory of rational option pricing. *Bell Journal of Economics and Management Science* 4: 141–183.

Montroll, E. W. and Weiss, G. H. (1965). Random walks on lattices. *Journal of Mathematical Physics* 6: 167–181.

Norros, I., Valkeila, E., Virtamo, J. et al. (1999). An elementary approach to a girsanov formula and other analytical results on fractional brownian motions. *Bernoulli* 5: 571–587.

Nourdin, I. (2012). *Selected Aspects of Fractional Brownian Motion, 4*. Springer.

Øksendal, B. (2003). *Stochastic Differential Equations*. Springer.

Osterrieder, D. et al. (2013). Interest Rates with Long Memory: A Generalized Affine Term-Structure Model. Tech. rep., School of Economics and Management, University of Aarhus.

Podlubny, I. (1998). *Fractional Differential Equations*. Academic Press.

Politi, M. and Scalas, E. (2007). Activity spectrum from waiting-time distribution. *Physica A: Statistical Mechanics and its Applications* 383: 43–48.

Politi, M. and Scalas, E. (2008). Fitting the empirical distribution of intertrade durations. *Physica A: Statistical Mechanics and its Applications* 387: 2025–2034.

Raberto, M., Scalas, E. and Mainardi, F. (2002). Waiting-times and returns in high-frequency financial data: an empirical study. *Physica A: Statistical Mechanics and its Applications* 314: 749–755.

Rachev, S. T., Kim, Y. S., Bianchi, M. L. and Fabozzi, F. J. (2011). *Financial Models with Lévy Processes and Volatility Clustering*. John Wiley & Sons.

Rogers, L. C. G. (1997). Arbitrage with fractional brownian motion. *Mathematical Finance* 7: 95–105.

Rostek, S. and Schöbel, R. (2013). A note on the use of fractional brownian motion for financial modeling. *Economic Modelling* 30: 30–35.

Samko, S., Kilbas, A. A. and Marichev, O. (1993). *Fractional Integrals and Derivatives*. CRC Press.

Samorodnitsky, G. (2007). Long range dependence. *Foundations and Trends® in Stochastic Systems* 1: 163–257.

Samorodnitsky, G. and Taqqu, M. S. (1994). *Stable Non-Gaussian Random Processes*. Chapman & Hall.

Sazuka, N., Inoue, J.-i. and Scalas, E. (2009). The distribution of first-passage times and durations in forex and future markets. *Physica A: Statistical Mechanics and its Applications* 388: 2839–2853.

Scalas, E. (2006). The application of continuous-time random walks in finance and economics. *Physica A: Statistical Mechanics and its Applications* 362: 225–239.

Scalas, E., Gorenflo, R., Luckock, H., Mainardi, F., Mantelli, M. and Raberto, M. (2004). Anomalous waiting times in high-frequency financial data. *Quantitative Finance* 4: 695–702.

Scalas, E., Gorenflo, R. and Mainardi, F. (2000). Fractional calculus and continuous-time finance. *Physica A: Statistical Mechanics and its Applications* 284: 376–384.

Scalas, E., Gorenflo, R. and Mainardi, F. (2004). Uncoupled continuous-time random walks: Solution and limiting behavior of the master equation. *Physical Review E* 69: 011107.

Schwab, C., Hilber, N. and Winter, C. (2007). Computational methods for quantitative finance. *Lecture notes and exercises, ETH Zurich*.

Shlesinger, M. and Montroll, E. (1984). On the wonderful world of random walks. in: J. leibowitz and E.W. montroll (eds.). *Nonequilibrium Phenomena II: From Stochastics to Hydrodynamics* : 1–121.

Shreve, S. E. (2004). *Stochastic calculus for finance II: Continuous-time models*, *11*. Springer Science & Business Media.

Teverovsky, V., Taqqu, M. S. and Willinger, W. (1999). A critical look at Lo's modified r/s statistic. *Journal of Statistical Planning and Inference* 80: 211–227.

Tsay, R. S. (2005). *Analysis of Financial Time Series*, *543*. John Wiley & Sons.

Wang, I. R., Wan, J. W. and Forsyth, P. A. (2007). Robust numerical valuation of European and American options under the CGMY process. *Journal of Computational Finance* 10: 31.

Wiersema, U. F. (2008). *Brownian Motion Calculus*. John Wiley & Sons.

Willinger, W., Taqqu, M. S. and Teverovsky, V. (1999). Stock market prices and long-range dependence. *Finance and Stochastics* 3: 1–13.

Index

Printed in the United States
By Bookmasters